Mohammad Hossein Keshavarz

Toxicity: 77 Must-Know Predictions of Organic Compounds

Also of interest

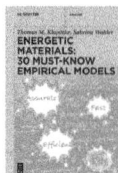

Energetic Materials: 30 Must-Know Empirical Models
Klapötke, Wahler, 2024
ISBN 978-3-11-109602-5, e-ISBN (PDF) 978-3-11-109702-2,
e-ISBN (EPUB) 978-3-11-109782-4

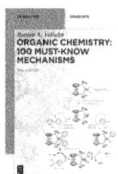

Organic Chemistry: 100 Must-Know Mechanisms
2nd Edition
Valiulin, 2023
ISBN 978-3-11-078682-8, e-ISBN (PDF) 978-3-11-078683-5,
e-ISBN (EPUB) 978-3-11-078701-6

Combustible Organic Materials.
Determination and Prediction of Combustion Properties
2nd Edition
Keshavarz, 2022
ISBN 978-3-11-078204-2, e-ISBN (PDF) 978-3-11-078213-4,
e-ISBN (EPUB) 978-3-11-078225-7

Energetic Materials Encyclopedia
Vol. 1-3
Klapötke, 2021
ISBN 978-3-11-067465-1

The Properties of Energetic Materials.
Sensitivity, Physical and Thermodynamic Properties
2nd Edition
Keshavarz, Klapötke, 2021
ISBN 978-3-11-074012-7, e-ISBN (PDF) 978-3-11-074015-8,
e-ISBN (EPUB) 978-3-11-074024-0

Energetic Compounds.
Methods for Prediction of Their Performance
2nd Edition
Keshavarz, Klapötke, 2020
ISBN 978-3-11-067764-5, e-ISBN (PDF) 978-3-11-067765-2,
e-ISBN (EPUB) 978-3-11-067775-1

Mohammad Hossein Keshavarz

Toxicity: 77 Must-Know Predictions of Organic Compounds

Including Ionic Liquids

DE GRUYTER

Author
Prof. Dr. Mohammad Hossein Keshavarz
Department of Chemistry
Malek-Ashtar University of Technology
Shahin-Shahr, Iran
keshavarz7@gmail.com

ISBN 978-3-11-118912-3
e-ISBN (PDF) 978-3-11-118967-3
e-ISBN (EPUB) 978-3-11-119092-1

Library of Congress Control Number: 2023939212

Bibliographic information published by the Deutsche Nationalbibliothek
The Deutsche Nationalbibliothek lists this publication in the Deutsche Nationalbibliografie;
detailed bibliographic data are available on the Internet at http://dnb.dnb.de.

© 2023 Walter de Gruyter GmbH, Berlin/Boston
Cover image: DuxX/iStock/Getty Images Plus
Typesetting: Integra Software Services Pvt. Ltd.
Printing and binding: CPI books GmbH, Leck

www.degruyter.com

Preface

Due to the advances in various methods for the prediction of toxicity of organic compounds and ionic liquids (ILs), it is necessary to review these methods for scientists and students. It is essential to compare the advantages and shortcomings of these methods. Since many organic compounds and ILs are synthesized each year, this book introduces suitable models for the assessment of their toxicities.

This book reviews the best predictive methods to assess the toxicity of organic compounds and ILs, which were derived by *in vitro* or *in vivo* experiments. Each chapter contains some complimentary problems with their answers. This book is of interest to any scientist or student working with new organic compounds and ILs or synthesized compounds. The introduced subjects are suitable for advanced students in chemistry, biochemistry, medicinal chemistry, and chemical engineering.

The introduced subjects are suitable for advanced students in chemistry, biochemistry, medicinal chemistry, and chemical engineering. Moreover, this book can be applied in the following fields:

1) It may be interesting for researchers including academics, national laboratories, and scientific agencies.
2) It can encourage R&D agencies to look for next-generation organic compounds and ILs.
3) It may be interesting for the general public due to the toxicity of many organic compounds and ILs.

This book contains five chapters. The first chapter provides needed basic knowledge about common organic solvents and their potential toxicities that will alert researchers to think twice and always think for their health as well as for the environment via safe and green practices. Since polycyclic aromatic hydrocarbons (PAHs) are organic compounds that are widely distributed in the air, water, and soil, this chapter describes traditional and current studies of PAH toxicities and the related bioaccumulation properties in aquatic animals. Organophosphates are one of the major constituents of herbicides, pesticides, insecticides, and nerve gas. This chapter reviews different classes of organophosphate pesticides, their environmental issues, analytical techniques for estimation, and eco-friendly biodegradation approaches for their efficient bioremediation. Some of the organic pollutants are carcinogenic and can induce serious health complications upon short- and long-term exposures. The adsorption of organic pollutants is one of the promising methodologies for their removal from the water. The second chapter reviews different predictive models for important classes of organic compounds, that is, nitroaromatics, aromatic aldehyde, aniline compounds, halogenated phenols, and so on. Each class will be discussed in one section. The third chapter discusses the best predictive correlations for important derivatives of organic compounds, that is, aromatic derivatives, phenol derivatives, benzene derivatives, and so on. Each derivative will be demonstrated in one section. The fourth chapter demonstrates general methods for a wide range of

https://doi.org/10.1515/9783111189673-202

aromatic and organic compounds. Each method will be discussed in one section. Finally, the fifth chapter expresses various reliable models for the assessment of the toxicity of ILs. Each approach will be illustrated in one section. Among different equations, bold 77 equations with "*" are important that correspond to title of book "77 Must-Know Predictions of Organic Compounds".

M. H. Keshavarz

Contents

Chapter 1
Toxicity Assessment

This chapter provides the needed basic knowledge about common organic solvents and their potential toxicities that will alert researchers to think twice and always think for their health as well as for the environment via safe and green practices. Since polycyclic aromatic hydrocarbons (PAHs) are organic compounds that are widely distributed in the air, water, and soil, this chapter describes traditional and current studies of PAH toxicities and the related bioaccumulation properties in aquatic animals. Organophosphates are one of the major constituents of herbicides, pesticides, insecticides, and nerve gas. This chapter reviews different classes of organophosphate pesticides, their environmental issues, analytical techniques for estimation, and eco-friendly biodegradation approaches for their efficient bioremediation. Some of the organic pollutants are carcinogenic and can induce serious health complications upon short- and long-term exposures. The adsorption of organic pollutants is one of the promising methodologies for their removal from water.

Toxicity of organic compounds may be expressed in terms of an oral LD_{50} dose (50% lethal dose), for example, in rats. Among protozoa, *Tetrahymena pyriformis* or *T. pyriformis* can also be used to study chemical toxicants over the years. Toxicity of ionic liquids (ILs) may be reported as effective nominal concentration EC_{50} (half-maximal effective concentration or the concentration of a drug inducing its half-maximal effective response), which is defined as the concentration of a desired IL that produces a mortality of 50% of the bacterial population. The toxicity of diverse IL subfamilies can be determined against different organisms' toxicity such as esterase enzyme, *Vibrio fischeri*, algae, and *Daphnia magna*.

1.1 Toxicity Measurements and Predictions

1.1.1 Dose Descriptors and In Silico Tools

Dose descriptors can identify the relationship between a specific effect of a compound and the dose at which it takes place. They are used to derive the no-effect threshold levels for human health (i.e., *DNEL* or reference dose (*RfD*)) and the environment (*PNEC*). It should be mentioned that DNEL is the level of exposure to a substance above which humans should not be exposed because the REACH (Registration, Evaluation, Authorization and Restriction of Chemicals) regulation defines *DNEL* as exposure levels beneath which a substance does not harm the human health [1]. The *PNEC* is also the concentration of a chemical that marks the limit at which no adverse effects of exposure in an ecosystem are measured. Dose of descriptors determine the hazards of the substance in terms of LD_{50}, LC_{50} (50% lethal concentration), *NOAEL* (no observed

https://doi.org/10.1515/9783111189673-001

adverse effect level), *NOAEC* (lower systemic toxicity or lower chronic toxicity), T_{25} (the chronic daily dose in mg per kg bodyweight which will give 25% of the animals' tumors at a specific tissue site), benchmark dose, EC_{50}, *NOEC* (no observed effect concentration), and so on. For example, LD_{50} is a statistically derived dose at which 50% of a large number of test animals of a particular species can be expected to die. Its value is expressed in milligrams of the substance being tested per kilogram of animal bodyweight (mg kg^{-1}). For inhalation toxicity, air concentrations are used for exposure values, that is, LC_{50}.

Experimental determination of the toxicological effects of chemicals is vital to assess their environmental monitoring and risk. Adverse reactions characterize acute toxicity, which can occur immediately or briefly after exposure to single or multiple doses of a chemical (within a 24-h period). This path represents one of the most common toxicity endpoints [2]. Rodent acute toxicity tests can assess the potential hazards of chemicals to human health [3]. It is usual to perform standard tests for the assessment of acute toxicity of chemicals through *in vivo* animal experiments. Animal experiments require intense labor and material inputs as well as complexity. It is very hard to rapidly determine acute toxicity exclusively using *in vivo* animal experiments alone because many chemicals are increasing dramatically [4]. Moreover, ensuring the safe use of chemicals is mandated for both importers and manufacturers to register toxicity information under the framework of the European Union (EU) REACH regulation [1]. Determination of the toxicity of a large number of chemicals via animal assays, considering the resources and animal welfare or ethics issues, is impossible [4]. Due to the variety and complexity of the toxicological pathways of chemicals, their toxicological mechanism is still the focus of active discussion, which need better theoretical methods. Thus, predictive methods such as *in silico* tools are used to the defendant and implement replacement, reduction, and refinement (the 3R principles of animal experiments) [4].

1.1.2 Aquatic Toxicity

For the environmental hazard and risk assessment of all types of chemicals, aquatic toxicity is of interest according to several pieces of EU regulation. Aquatic toxicity can determine the effect that chemicals exert on water organisms. It can be estimated by testing on different species: (i) plants or algae, (ii) invertebrates (e.g., *Daphnia magna* and *Vibrio fischeri*), and (iii) vertebrates (fish).

Toxicity can be given in terms of different values, such as EC_{50} or LC_{50}, lowest observed effect concentration, and *NOEC*. Toxicologists use the EC_{50} value to represent the effective concentration of material that causes 50% of the maximum response. For other responses, the symbol EC_x can also be used corresponding to the effective concentration associated with the $x\%$ response. The chemicals can be classified based on the Globally Harmonized System of Classification and Labeling of Chemicals [5]:

a) If $EC_{50} \leq 1$ mg L^{-1} for a chemical, it is highly toxic.
b) If a chemical has toxicity in the range of 1 mg L$^{-1} < EC_{50} \leq 10$ mg L^{-1}, it has moderate toxicity.
c) For a chemical containing poorly toxic, its toxicity is in the range of 10 mg L$^{-1} < EC_{50} \leq 100$ mg L^{-1}.

The fourth category can also be considered for nontoxic chemicals where they have $EC_{50} > 100$ mg L^{-1} [5]. Acute and chronic toxicities are two types of endpoints in aquatic toxicity. Acute toxicity represents the effect of the short-term exposure of a chemical to an organism. In contrast to acute toxicity, chronic toxicity is less common. Since chronic toxicity describes the effect after long-term exposure, it is needed if the result of the acute toxicity tests indicates a risk or when long-term exposure is expected. For some pharmaceuticals such as antibiotics, chronic toxicity is more appropriate than acute because they are highly hydrophilic and ionized. Since they need a certain period to enter the cell membranes and exert their full toxic effects, their bioactivity cannot reach equilibrium through short-term exposure. Thus, their short-term toxicity is lower as compared to long-term exposure. For determining the toxicity of pharmaceuticals, the mode of action of some compounds is closely related to the exposure time. This situation makes chronic assays more appropriate than acute assays [6].

There are five modes of action to help explain the toxicity effect of a chemical within a certain chemical class on a tested organism [5, 7]:
(i) *Inert chemicals (baseline toxicity):* This mode can be applied for chemicals acting by nonpolar narcosis. Since they are not reactive when considering overall acute effects, they do not interact with specific receptors in an organism. The term "narcosis" is used for the mode of action of such compounds. Narcosis is a nonspecific mode of action, which depends on the hydrophobicity of the compound.
(ii) *Less inert chemicals:* This mode is used for compounds acting by polar narcosis containing phenols and anilines. These chemicals may have hydrogen bond donor acidity.
(iii) *Reactive chemicals:* This mode can be used for chemicals including aldehydes and epoxides with nonspecific reactivity. Chemicals of all kinds of different modes of action in this class with enhanced toxicity are related to the unselective reaction with some chemical structures.
(iv) *Specifically acting chemicals:* This class is used for chemicals with specific reactivity, which includes pesticides, some PAH metabolites, and polychlorinated biphenyls. The compounds of this category may have a specific interaction with certain receptor molecules.
(v) *Those chemicals are not covered by (i)–(iv) classification.*

It is important to know the mode of toxic action for the application of certain toxicity models.

1.2 Quantitative Structure–Activity/Property/Toxicity Relationships (QSAR QSPR/QSTR)/

Quantitative structure–activity/property/toxicity relationships (QSAR/QSPR/QSTR) are very helpful in predicting the biological activity, property, and toxicity of a given set of molecules as a linear combination of certain descriptors. Since QSAR connects the structure of a molecule to those properties that show medicinal applications, it is considered as an activity of a molecule. QSAR models can design molecules to serve as a better drug or as a better substance in terms of economy, utility, durability, and environmental effects. QSPR models can predict the observed macroscopic properties of systems such as various physical, chemical, biological, and technological properties through structural alteration changes. QSTR models can predict the toxicological effects of molecules through molecular structures. They can estimate the toxicity of molecules that have the potential to act as toxic substances. They help in predicting the reactivity of an unknown molecule based on the behavior of an analogous set of chemically or structurally similar compounds. QSTR models are built on the same premise as QSAR/QSPR/QSTR models where they usually use molecular descriptors. The molecular descriptor includes theoretical and experimental parameters such as the number of atoms, chemical MLI (molecular connectivity index), and bond numbers. QSAR/QSPR/QSTR approaches can predict various properties within compounds and a series of compounds [8], where they may need complex descriptors, special computer codes, and expert users as well as the presence of similarity between molecular structures of the compounds. Multiple linear regression (MLR), principal component analysis, genetic algorithm, artificial neural network, partial least-squares method, and support vector machine are common mathematical modeling algorithms that are favorable to process linear or nonlinear relationships [9]. The selection of molecular descriptors can determine the quality of models because more descriptors may result in overfitting. Moreover, the selected molecular descriptors should be mutually independent or less correlated.

Many QSAR/QSPR/QSTR investigations have been made using MLR, which is a commonly used method in QSAR due to its simplicity, transparency, reproducibility, and easy interpretability. The generalized expression of an MLR equation is given as follows:

$$Y = a_0 + \sum_{i=1}^{n} a_i X_i \tag{1.1}$$

where Y is the dependent or response variable, X_i are features or independent variables (descriptors) present in the model with the corresponding regression coefficients a_i, and a_0 is the constant term of the model. The descriptors present in an MLR model should not be much intercorrelated as well as the number of observations and descriptors should bear a ratio of at least 5:1 [10]. The correlation matrix of the used descriptors can assess the QSAR/QSPR/QSTR model quality. It verifies the degree of correlation among each couple of modeling descriptors in a QSAR/QSPR/QSTR model [11].

1.3 Statistical Parameters for Assessment of QSAR/QSPR/QSTR Models

Internal and external validations can assess the performances of QSAR/QSPR/QSTR models because they ensure their robustness. The parameters of the coefficient of determination (r^2), the mean absolute error, and root mean square error evaluate the goodness of fit between predicted and the experimental data. Cross-validation or internal validation shows different proportions of compounds, which are iteratively held out from the training set for evaluation of QSAR/QSPR/QSTR models. The parameter q^2 is commonly a mean cross-validated r^2. For the validation of a QSAR/QSPR/QSTR model, the q^2 coefficients of leave-one-out (LOO) or q^2_{LOO} and leave-many-out (LMO) or q^2_{LMO} are usually used [12, 13].

For the training set data, the Y-scrambling or randomization test can confirm a robust QSAR/QSPR/QSTR model [14]. It expresses the chance of the relations between descriptors [15]. It produces the dependent variable of MLR models while keeping the descriptors unchanged. The developed QSAR/QSPR/QSTR models are robust when the resulting r^2_{YS} and q^2_{YS} values for the Y-scrambling test of QSAR/QSPR/QSTR models are significantly low (close to 0) for several trials.

The external validation method can evaluate a QSAR/QSPR/QSTR model through a test set of chemicals, which may assess its predictive power [16]. The effects of the composition and the size of the training or validation (test) set for predicting the power of a QSAR/QSPR/QSTR model have been illustrated elsewhere [17–20]. Statistical external validation evaluates the estimating power of the methodologies [17–21]. It is desirable to partition the measured data into training and test sets to confirm a QSAR/QSPR/QSTR model. The test set displays the domain of applicability of the model in terms of property values. The QSARINS software divides the data set randomly into training and test sets [11, 22].

Several additional external validation coefficients of a QSAR/QSPR/QSTR model are also performed on the test molecules as q^2_{F1} [23], q^2_{F2} [24], q^2_{F3} [25] and concordance correlation coefficient (CCC) [25–27]. Three parameters such as r^2_m (test), r^2_m (overall), and Δr^2_m (test) were described elsewhere [14, 28, 29]. Threshold values for q^2_{F1} [23], q^2_{F2} [24], and q^2_{F3} [25] are more than 0.5 except Δr^2_m and CCC which are less than 0.2 and more than 0.8 [30], respectively.

The applicability domain (AD) can characterize a QSAR/QSPR/QSTR model for a small amount of data [31]. It is defined based on the molecules of the training set for a QSAR/QSPR/QSTR model. If the leverage value (h) [32] is less than the critical value (h^*) for a chemical of the training set, the predicted result is consistent with the experimental value. If the structure of a chemical is far off from the training set, it can be considered outside the AD of the model. The predicted results of the test set are unreliable for those compounds containing $h > h^*$. The critical value is given as $h^* = 3(d + 1)/n$, where d and n are the number of model variables and some molecules in a work set [33]. QSARINS software [11, 22] can visualize the AD of a new model. Williams plot [34, 35] shows the plot of standardized cross-validated residuals versus h values. It can detect the compounds with cross-validated standardized residuals of more than 3 standard deviation units ($>3\sigma$) or the response outliers, and $h > h^*$ [36]. If the training set chemicals with high values of h do not fit a QSAR/QSPR/QSTR model, they are called bad "high leverage points" (bad influence points), which can destabilize the model [37]. If the chemicals with high leverage points fit a QSAR/QSPR/QSTR model well, they are called "good high leverage points," which can stabilize the model.

The Insubria graph provides the plot of the estimated values versus h without using the experimental data [11, 22]. The term "%AD" can estimate the percentage of reliable data because it shows the percentage of chemicals given into the structural domain of each full model. Insubria graphs confirmed that a QSAR/QSPR/QSTR model with a high value of AD% can choose as a reliable prediction zone by an arbitrary cutoff [30].

1.4 Common Organic Solvents and Their Toxicities

Organic solvents are used to dissolve other chemicals such as polymers where solvent selection for polymers has wide applications in industries to recycle degraded polymers with environmentally compatible by-products [38]. It is important to avoid phase separation during the synthesis of polymers as well as to improve stable formulations with high-quality standards for entering the environment [39]. Since it is essential to understand the reactivity of respective solvents on specific reaction conditions, the selection of appropriate solvents for a reaction is to be noted always. For chemical reactions, the choice of solvents is usually done through the experience of chemists or engineers with particular solvents or the following similar pattern literature review and practice it in a laboratory setting [38].

Since polymers have wide applications in life, reliable prediction of the solubility of polymers in environmentally compatible solvents is very important. Solubility parameters can select compatible solvents for predicting the vapor pressure of the solvent in polymer solutions for reaction systems and the removal of volatile substances [40], forecasting phase equilibria for polymer–binary [41], random copolymer [42], polymer–polymer [43], and multicomponent solvents [44]. The solubility parameter quantifies the interactions between polymers and solvents. The cohesive energy density provides the energy required to break all intermolecular physical links in a unit volume of the material [45].

The common organic solvents are usually aliphatic hydrocarbons, aromatic hydrocarbons, cyclic hydrocarbons, halogenated hydrocarbons, amines, ketones, esters, ether, aldehyde, and alcohols as well as organic compounds containing specific polar groups such as dimethylsulfoxide. Since there are a wide range of different classes of organic compounds, it is suitable to select those solvents with lesser toxicities.

1.4.1 Section of Solvents for Polymers with Less Toxicity

The solubility parameters are related to three types of cohesive energies, which include dispersion, polar, and hydrogen bonding forces [12]. Dispersion forces depend on the polarizabilities of neighboring molecules because the temporary random dipole of a molecule relates with the dipoles induced in all neighboring molecules [13]. Polar cohesive forces depend on the permanent dipole moment and the interactions between different structural groups [14]. Hydrogen bonding provides forces much stronger than ordinary dipole–dipole interactions [16].

The Hildebrand and Hansen solubility parameters express solvent–polymer compatibility [46]. The Hildebrand solubility parameter (δ) is the square root of the cohesive energy density of a substance, which shows a good solvent or nonsolvent for a polymer [46]. Good solvents contain δ values within ±2 MPa$^{1/2}$ of the polymer δ value [47]. The Hansen solubility parameters represent the dispersion (δ_D), polar (δ_P), and hydrogen bonding (δ_H) components of the Hildebrand parameter δ as follows [48]:

$$\delta^2 = \delta_D^2 + \delta_P^2 + \delta_H^2 \tag{1.2}$$

The parameters δ^2, δ_D^2, δ_P^2, and δ_H^2 are interrelated to the total, dispersion, polar, and hydrogen bonding cohesive energies.

Solvation of a polymer lowers the Gibbs free energy of the system to dominate the solvent–solvent and polymer–polymer interactions through strong interactions between a solvent and the polymer [49]. Crystallinity, cross-linking, and increasing molecular weight can decrease the solubility of a polymer [50]. Specific interactions occur between the polymer and the solvent, or among the two blended polymers by the solvation of polymers, pigments, plasticizers, and other additives [51]. Similar interactions have a critical role in solvation or miscibility [52].

The solvent selection practice is important in increasing the concern about safety, effectiveness, and available alternative. Green chemistry influences the organic chemist to search for possible less toxic organic solvents. Products with minimal toxicity, minimal use, and reuse options are essential for the big pharmaceutical company to develop their solvent selection guide [53]. The solvent selection guide by Pfizer company is given in Table 1.1 [54]. The solvents are subjected to the red category due to having these potencies: toxic, carcinogenic, mutagenic, low flash point, environmental risk of ozone layer depletion, and so on.

1.4.2 The Use of a Simple Approach for Selection Solvents of Polymers

The parameters δ^2, δ_D^2, δ_P^2, and δ_H^2 have been reported only for a relatively small number of materials [50]. Group contribution methods and QSPR methodology can be used to predict the solubility parameter δ [55–59]. Group contribution methods use

Table 1.1: The solvent selection guide is given by Pfizer company [54].

Preferred	δ (MPa$^{1/2}$)	Usable	δ (MPa$^{1/2}$)	Undesirable	δ (MPa$^{1/2}$)
Acetone	20.3	Acetonitrile	24.3	Benzene	18.8
1-Butanol	23.3	Acetic acid	20.7	Chloroform	19.0
t-Butanol	21.7	Cyclohexane	16.8	Carbon tetrachloride	17.6
Ethyl acetate	18.6	Dimethylsulfoxide	29.7	Dichloromethane	–
Ethanol	–	Ethylene glycol	29.9	Diethyl ether	15.1
Isopropyl acetate	–	Heptane	15.1	Dichloroethane	–
2-propanol	23.5	Isooctane	–	Di-isopropyl ether	14.1
1-Propanol	24.3	Methylcyclohexane	19.0	Dimethylformamide	24.8
Methanol	29.7	2-Methyl THF	–	Dioxane	20.5
Methyl ethyl ketone	19.0	Methyl t-butyl ether	–	Dimethyl acetate	–
Water	47.9	Toluene	18.2	Dimethoxyethane	–
		Tetrahydrofuran	18.6	Hexane(s)	14.1
		Xylenes	18.0	N-Methylpyrrolidinone	–
				Pentane	23.5

The reported solubility parameters [50] of solvents are also given (MPa$^{1/2}$).

additive contributions from a variety of atoms, and polar and nonpolar groups but they cannot account for the presence of neighboring groups or conformational influences, and those polymers having a specific functional group where it is missed from the database of functional groups [56]. Several QSPR models were used for the prediction of the solubility parameter δ of polymers but they need a set of complex molecular descriptors from polymeric repeating unit structures $-(C^1H_2-C^2R^3R^4)-$, which need specific computer codes and expert users [55, 57–59]. A simple approach has been introduced to find the value of δ for a desired polymer only from elemental composition and structural parameters of repeating unit structures $-(C^1H_2-C^2R^3R^4)-$ as follows [60]:

$$\delta = 22.96 - \frac{45.16n_H - 300.90n_N - 51.06n_O - 129.21n_S + 243.63n_F + 287.71n_{Si}}{Mw_{unit}}$$
$$+ 3.75\delta^{Inc} - 2.04\delta^{Dec} \tag{1.3*}$$

where δ is given in MPa$^{1/2}$; n_H, n_N, n_O, n_S, n_F, and n_{Si} are the number of moles of hydrogen, nitrogen, oxygen, sulfur, fluorine, and silicon atoms; Mw$_{unit}$ is the molecular weight of repeating unit structure in g/cm^3, δ^{Inc} and δ^{Dec} are two correcting functions for adjustment of the underestimating and overestimating results from the contribution of elemental composition as compared to experimental data. Different values of δ^{Inc} and δ^{Dec} are given as follows:

(i) *Hydrogen bonding groups:* The values of δ^{Inc} are 3.0, 2.0, and 1.25 for the existence of more than one –OH, one –OH (without the existence of more than one polar group), and phenolic hydroxyl group, respectively. The value of $\delta^{Inc} = 0.0$ for the presence of more than one polar group. For example, the values of δ^{Inc} are 3.0, 2.0, and 0.0

for cellulose, poly(vinyl alcohol), and cellulose diacetate, respectively. The δ^{Inc} value is 1.0 for the existence of –COOH or –CONH$_2$. The value of δ^{Inc} = 0.25 for the attachment of the alkoxy group to the main chain containing the alkyl group substituent. The δ^{Inc} value is 5.0 for the presence of –S(O)$_2$–NH$_2$. The δ^{Inc} value for the other hydrogen bonding groups can be neglected, that is, δ^{Inc} = 0.0.

(ii) *Polar groups:* The δ^{Inc} value is 1.0 for the presence of –NH-(CH$_2$)$_n$-C(=O)O–, –NH-(CH$_2$)$_n$-NH–, –CH$_2$-CF$_2$–, and –CF$_2$-S– in the main chain. The values of δ^{Inc} are 2.0 and 0.8 for the existence of the furan ring in the main chain and the attachment of –C(=O)CH$_2$– to the main chain, respectively. The δ^{Inc} value is 1.0 for the attachment of –CN to the main chain without the alkyl group.

The δ^{Dec} value is 1.0 for the existence of CF$_3$-CF$_2$–, (CH$_3$)$_3$-O(C = O)–, and –CH$_2$-O(C = O)-CH$_3$ as well as the attachment of –CN to the repeating unit structure containing alkyl group and without any polar group. The value of δ^{Dec} = 1.5 for the existence of –CH$_2$-S– in the main chain. The δ^{Dec} value is 0.6 for the attachment of the phenyl group, which may contain an electron-withdrawing group, to the repeating unit structure without other substituents. The value of δ^{Dec} is 2.0 for the existence of –NHC(=O)NH– in the main chain.

Example 1.1: Polymer poly-AMMO has the following repeating unit structure. It can be used as an energetic binder in polymer-bonded explosives [61]. Find a good solvent for this polymer.

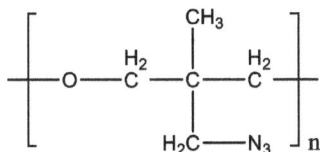

Answer: The use of eq. (**1.3***) gives

$$\delta = 22.96 - \frac{45.16n_H - 300.90n_N - 51.06n_O - 129.21n_S + 243.63n_F + 287.71n_{Si}}{Mw_{unit}} + 3.75\delta^{Inc} - 2.04\delta^{Dec}$$

$$\delta = 22.96 - \frac{45.16 \times 9 - 300.90 \times 3 - 51.06 \times 1 - 129.21 \times 0 + 243.63 \times 0 + 287.71 \times 0}{5 \times 12.011 + 9 \times 1.008 + 3 \times 14.007} + 3.75 \times 0 - 2.04 \times 0$$

$$= 27.80 \text{ MPa}^{1/2}$$

Thus, suitable solvents are in the range of δ = 25.8–29.8 MPa$^{1/2}$.

Equation (**1.3***) provides a suitable pathway to select suitable solvents with less toxicity. Solvents with similar δ sometimes do not show the same behavior against certain polymers. Poorly, moderately, and strongly hydrogen bonding solvents can be selected suitably using eq. (**1.3***) based on the existence of hydrogen bonding and polar groups in the repeating unit structure of a polymer because δ_D of a solvent is equal to that of a nonpolar substance. In contrast to available QSPR methods, eq. (**1.3***) can be easily applied for amorphous polymers and the amorphous phases of semicrystalline polymers. Equation (**1.3***) cannot be applied to polymers of very high crystallinity because they are insoluble in solvents whose solubility parameters perfectly match their own.

1.4.3 General Comments on Using Solvents

If the skeleton of organic solvents contains carbon, hydrogen, and nitrogen as major atoms without polar groups, they have high lipophilicity and volatility. Since the lipophilicity of solvents influences their distribution to various body parts, it is essential for converting them to a water-soluble form via several osmotic conversions that enhance the excretion via the kidney. Since organic solvents have a high lipophilicity character, they may arrive in the brain and affects severely. Organic solvents are usually highly volatile and mix with air rapidly upon exposure to air. Since they can enter the lungs via respiration, lungs are alerted to have enough ventilation in the workplace. Chemists and chemical engineers work in a lab or chemical industries for a long time, where they may handle various kinds of toxic reagents and organic solvents almost every day.

The extraction of organic compounds and excretion of undesired side products to the water layer depend on the hydrophobic nature of the organic solvents and the polar water. Incineration and distillation processes may be used to prevent entering waste solvents into the environment. For chemical and pharmaceutical industries, the separation action plays a critical role in drug development research institutes. It is required to lose liters of the organic solvents for getting a milligram scale of the compound where solvent accounts for 40–70% of the overall research cost [62].

Due to the lipophilic nature of organic solvents, their absorption occurs immediately after inhalation or surface dermal contact, or oral exposure [63]. The metabolism and long-term or shorter deposition is affected once the solvent is absorbed through the route of exposure and the chemical–physical nature of that solvent. Since the metabolism and excretion can occur immediately with the liver and lungs, the relative toxic metabolite of solvents depends on the individual chemical nature. Some solvent molecules are metabolized to less toxic but some of them to a severely toxic metabolite. For the unmetabolized solvents, distributions occur largely in fatty tissue that affects the human body on a long-term basis.

Material Safety Data Sheet provides the proper knowledge about the individual chemicals such as solvents including potential hazards, safety measures, and other handling procedures. It is recommended to use a fully functional chemical fume hood for the handling of any chemicals. For the contact of chemicals to the eye, remove any contact lenses and immediately flush with running water for at least 15 min. It is essential to keep the eyelid open, which allows proper air supply to the eye for enhancing the fast evaporation of chemicals from the eye surface and decreasing further risk. For the contact of chemicals with the skin, immediately wash with an excess of water, cover the irritated skin with an emollient, remove contaminated clothes and shoes, and get medical attention in all cases if needed.

1.5 Toxicities of Polycyclic Aromatic Hydrocarbons

PAHs contain two or more condensed aromatic rings, for example, naphthalene and anthracene. Since PAHs are ubiquitous compounds in air, water, and soil, they are categorized as general environmentally harmful pollutants. PAHs are widely distributed in the aquatic environment in four types [64]:
1. Petrogenic – derived from fuels
2. Pyrogenic – derived from an incomplete combustion process
3. Biogenic – generated by organic metabolism
4. Diagenetic – generated by the transformation process in sediment

Petrogenic and pyrogenic sources are important contributors to environmental PAH pollution in aquatic ecosystems.

Oil spill accidents are among the most concerning exposure events, which introduce major components of crude oil including PAHs, aliphatic saturated hydrocarbons, aliphatic unsaturated hydrocarbons, and alicyclic saturated hydrocarbons into the environment [65]. Among these hydrocarbon chemicals, specific toxicity on the ecosystem from PAHs is especially concerning. Crude oil containing PAHs has toxic effects, for example, immunotoxicity, embryonic abnormalities, and cardiotoxicity, for wildlife including fish, benthic organisms, and marine vertebrates [66].

1.5.1 Carcinogenicity of PAHs

Carcinogenicity of PAHs is the most concerning toxicity upon transporting into cells because of their hydrophobicity [67], which induces gene expression of the cytochrome P450 (CYP) enzyme group [68]. Since CYP enzymes metabolize PAHs into additional metabolites, several intermediates in this metabolic pathway can bind to DNA, which becomes mutagenic/carcinogenic. The International Agency for Research on Cancer classified benzo[a]anthracene, benzo[a]pyrene, and dibenz[a,h]anthracene as being probably carcinogenic chemicals (group 2A). Moreover, the United States Environmental Protection Agency (US EPA), besides the 3 mentioned PAHs, considered further 13 PAHs carcinogenic chemicals and they are important organic pollutants in the environment and human society, which include naphthalene, acenaphthylene, acenaphthene, fluorene, phenanthrene, anthracene, fluoranthene, pyrene, chrysene, benzo[*b*]fluoranthene, benzo[*k*]fluoranthene, indeno[1,2,3-*c,d*]pyrene, and benzo[*g,h,i*]perylene. It was indicated that there are other types of toxicities from PAHs, which contain developmental toxicity, genotoxicity, immunotoxicity, oxidative stress, and endocrine disruption [69].

1.5.2 Octanol/Water Partition Coefficient (K_{OW})

The n-octanol/water partition coefficient (K_{OW}) represents the distribution of a chemical's concentration in octanol and water when the octanol–water system is at equilibrium [70]. It can be used to evaluate the hydrophobicity, toxicity, or activity of compounds. It can investigate the features of solvents during extraction, for environmental risk assessment, and for drug design in various fields [71]. It can estimate the fate of a wide extent of chemicals as it is indeed associated with the hydrophilic–lipophilic balance of a substance [72]. Pollutants with low values of K_{OW} < 10 and a low value of Henry's law constant are usually considered hydrophilic. In contrast, pollutants having K_{OW} values >104 are generally regarded as hydrophobic. Thus, pollutants could be adsorbed on sediments or soil, which leads to an increased chance of bioaccumulation [73]. K_{OW} can be used to interpret the biological activities of certain substances, especially in pharmaceuticals, in which a partial model of hydrophobic constituent existing in the receptor site is characterized by the octanol phase [74]. The measurement of K_{OW} is typically described in logarithmic form as log K_{OW}, which can be implemented in lab handling, but the existing data remain limited for numerous organic chemicals. Since the experimental determination methods for K_{OW} are labor-intensive, time- and resource-consuming, some methods have been developed for the prediction of log K_{OW} [75–78].

K_{OW} of each PAH congener, concentration in environmental media, bioavailability, and depuration/excretion of PAHs are important factors for the bioaccumulation of PAHs in aquatic animals [79]. Since PAHs are hydrophobic chemicals that have a high affinity with organic matter in water and sediment, they have high K_{OW} values, which are more predominant in high-molecular-weight PAHs (more than five-ring). High-molecular-weight PAHs have the same trend with typical persistent organic pollutants, such as polychlorinated biphenyls, to provide high K_{OW} values, which suggest a high bioaccumulation factor. This situation is rarely observed in several trophic biomagnification studies [80]. Moreover, species differences in the metabolism capacity of PAHs are strongly suggested for fish and invertebrates [81], which may be caused by species differences in the intake pathway and efficiency, the capacity of xenobiotics to metabolize, and the ability of depuration/excretion.

1.6 Organophosphate Pesticides and Their Toxicities

Organophosphate pesticides are the ester forms of phosphoric acid including thiol or amide derivatives of thiophosphoric, phosphinic, phosphonic, and phosphoric acids with additional side chains of phenoxy, cyanide, and thiocyanate groups, which are usually considered as safe for agriculture use due to their relatively fast degradation rates [82]. They are the main components of herbicides, pesticides, and insecticides as well as the main components of nerve gas [83]. Since they contain C–P linkage as a group of biogenic and synthetic compounds, they are resistant to thermal hydrolysis, photolytic degradation, and chemical decomposition as compared to similar compounds containing more reactive S–P, O–P, or N–P linkages [84]. Organophosphate pesticides usually contain a phosphoryl group, two lipophilic groups bonded to the phosphorus, and a leaving group bonded to the phosphorus which is often a halide [85].

Different levels of toxicity in animals, humans, plants, and insects may be produced through acute or chronic exposure to organophosphate pesticides. Since most organophosphate pesticides inhibit acetylcholinesterase activity, they can affect the nervous system in both aquatic and terrestrial fauna [86]. They can also produce neuroteratogenicity and genotoxicity, which include ecological and adverse environmental impact [87].

Organophosphate pesticides have a risk of metabolic, endocrine, neurological, hepatorenal disorders, neuritis, and psychiatric manifestations [82]. They may increase bladder cancer and leukemia in farmers followed by genotoxic effects. Table 1.2 shows the values of toxicity and log K_{OW} of common organophosphate compounds [83].

Table 1.2: The values of toxicity and log K_{OW} of common organophosphates [83].

Name	CAS no.	Toxicity	log K_{OW}
Acephate	30560-19-1	Acute toxicity; LD_{50} oral – rat – 700 mg/kg; dermal – LD_{50} rabbit – 2.000 mg/kg	−0.90
Azinphos methyl	86-50-0	Acute toxicity; LD_{50} oral – rat – 7 mg/kg; LC_{50} inhalation – rat – 1 h – 69 mg/kg	2.53
Bensulide	741-58-2	LD_{50} oral – rat – 271 mg/kg; LD_{50} dermal – rabbit –2.000 mg/kg	4.12
Bilanafos	741-58-2	Acute toxicity; LD_{50} oral – rat – 275 mg/kg; LD_{50} dermal – rabbit – 2.000 mg/kg	3.96
Chlorpyrifos methyl	2921-88-2	Acute toxicity; LD_{50} oral – rat – 1.828 mg/kg; LC_{50} toxicity – rabbit – >2.000 mg/kg.	4.31
Cyanofenphos	13067-93-1	Toxicity to fish; LC_{50} – Poecilia sp. – 0.66 mg/L	4.20
Diazinon	333-41-5	Acute toxicity; LD_{50} oral – rat – 1.012 mg/kg; LD_{50} oral – rat – 696 mg/kg; LC_{50} inhalation – rat – 4 h – toxicity >5.400 mg/kg; LD_{50} dermal – rabbit – >2.020 mg/kg	3.86
Ethion	63-12-2	Acute toxicity; LD_{50} oral – rat – 13 mg/kg; LC_{50} inhalation – rat – 864 mg/kg; LD_{50} dermal – rat – 62 mg/kg	5.07
Etrimfos	38260-54-7	Acute toxicity; LD_{50} oral – rat – 1.800 mg/kg; LD_{50} dermal – rabbit – >500 mg/kg	2.94
Fosamine	59682-52-9	Acute toxicity; LD_{50} oral – rat – 11.000 mg/kg; LC_{50} inhalation – rat – 1 h – >57.570 mg/kg; LD_{50} dermal – rabbit – >1.660 mg/kg	−2.65
Glufosinate	51276-47-2	Acute toxicity; LD_{50} oral – rat – 1.620 mg/kg; LC_{50} inhalation – rat – 4 h – 1.260 mg/m^3; LD_{50} dermal – rat – >2.000 mg/kg	−3.96

Table 1.2 (continued)

Name	CAS no.	Toxicity	log K_{ow}
Glyphosate	1071-836	Acute toxicity; LD_{50} oral – rat – 5.000 mg/kg; LD_{50} dermal – rabbit – 5.000 mg/kg; LD_{50} intraperitoneal – rat – 235 mg/kg	−4.47
Isoxathion	18854-01-8		3.73
Methamidophos	10265-92-6	Acute toxicity; LD_{50} oral – rat – 7.5 mg/kg; LC_{50} Inhalation – rat – 4 h – 162 mg/kg; LD_{50} dermal – rabbit – 100 mg/kg	−0.82
Methidathion	18854-01-8	Acute toxicity; LD_{50} oral – rat – 20 mg/kg; LC_{50} inhalation – rat – 4 h – 50 mg/m^3; LD_{50} dermal – rabbit – 196 mg/kg	2.21
Methamidophos	10265-92-6	Acute toxicity; LD_{50} dermal – rabbit – 270 mg/kg	−0.35
Omethoate	1113-02-6	Acute toxicity; LD_{50} oral – rat – 30 mg/kg; LD_{50} inhalation – mouse – 4 h – 140 mg/m^3; LC_{50} inhalation – rat – 1 h – >1.500 mg/m^3; LD_{50} dermal – rat – 700 mg/kg; LD_{50} intraperitoneal – rat −14,400 mg/kg	−0.74
Phenthoate	2597-03-7	Acute toxicity; LD_{50} oral – rat – 71 mg/kg; LC_{50} inhalation – rat – 4 h – 59 mg/kg; LD_{50} dermal – rat – 700 mg/kg	3.54
Phorate	298-02-2	Acute toxicity; LC_{50} inhalation – rat – 1 h – 11 mg/kg	3.56
Phoxim	14816-18-3	Acute toxicity; LD_{50} oral – rat – 300 mg/kg; LC_{50} inhalation – rat – 1 h – >3.200 mg/kg; LD_{50} dermal – rat – 1.000 mg/kg	4.39
Quinalphos	13593-03-8	Acute toxicity; LD_{50} oral – rat – 26 mg/kg; LC_{50} inhalation – mouse – 4 h – 330 mg/kg; LD_{50} dermal – rat – 300 mg/kg	4.45
Sulfotep	3689-24-51	Acute toxicity; LD_{50} oral – rat – 5 mg/kg; LC_{50} inhalation – rat – 4 h – 38 mg/m^3; LD_{50} dermal – rabbit – 20 mg/kg	3.98
Terbufos	13071-79-9	Acute toxicity; LD_{50} oral – rat – 1.7 mg/kg; LD_{50} dermal – rat – 9.5 mg/kg	4.56
Thiometon	640-15-3	Acute toxicity; LD_{50} oral – rat – 227 mg/kg; LC_{50} inhalation – rat – 60 mg/kg; LD_{50} dermal – rat – 1.100 mg/kg	3.15
Thionazin	297-97-2	Acute toxicity; LD_{50} oral – rat – 5 mg/kg; LD_{50} dermal – rat – 3.5 mg/kg	1.86
Triazophos	24017-47-8	Acute toxicity; LD_{50} oral – rat – 57 mg/kg; LC_{50} inhalation – rat – 4 h – 280 mg/m^3; LD_{50} dermal – rat – 1.100 mg/kg	3.34
Vamidothion	2275-23-2	Acute toxicity; LD_{50} oral – mouse – 300 mg/kg; LD_{50} oral – rabbit – 3.200 mg/kg	0.23

1.7 Prediction of Henry's Law Constant of Pesticides, Solvents, Aromatic Hydrocarbons, and Persistent Pollutants

Henry's law indicates that the quantity of dissolved gas in a liquid is proportional to its partial pressure above the liquid. Henry's law constant is the proportionality constant which shows a partition coefficient. It has wide applications in the environment and chemical industries such as (i) the botanical biofiltration of volatile organic compounds with active airflow [88], and (ii) the assessment of Henry's law constant for some organic compounds as a function of temperature and water composition [89]. It shows the tendency of a compound to volatilize from the water surface into the atmosphere. Thus, it is a worthy indicator of a compound's volatility because its higher value for the chemical is more probable to volatilize it from aqueous solutions [90]. Accurate knowledge of Henry's law constant can help scientists to study the environmental sciences because it can demonstrate the movement of different types of organic compounds inside and outside aquatic ecosystems [91]. Due to the adsorption of small amounts of solute on the wall of the device and analytical detection limits of low concentrations of highly hydrophobic compounds, it is hard to measure Henry's law constant accurately [92]. The experimental data of Henry's law constant are available only for a relatively small number of chemicals in the literature because its measurement is expensive [93]. The measured values of Henry's law constant values at 25 °C in terms of atm m^3 mol^{-1} for different types of solvents, pesticides, persistent pollutants, and aromatic hydrocarbons were available in the HENRYWIN module of the EPI Suite [94].

Some attempts containing indirect and direct approaches have been done to introduce reliable methods for the prediction of Henry's law constant of a wide range of pure chemicals. Indirect methods have been developed to estimate Henry's law constant of different types of compounds in the water where they use some correlations between Henry's law constant and the other vapor–liquid equilibrium data such as activity coefficient, vapor pressure, and aqueous solubility [95]. These approaches require highly accurate vapor–liquid equilibrium data of the desired physical properties to obtain highly accurate values of Henry's law constant. Moreover, they cannot be applied to estimate Henry's law constant of those chemicals if the requested physical properties are not available. Direct methods usually use QSPR modeling to predict Henry's law constant but many of them can estimate Henry's law constant in homologous sets of chemicals [96–103]. General approaches to Henry's law constant parameter for different classes of organic compounds are limited because they usually require three-dimensional molecular descriptors. Duchowicz et al. [103] introduced a QSPR model based on complex descriptors for different types of organic compounds including pesticides, solvents, persistent pollutants, and aromatic hydrocarbons as follows:

$$\log k_H = -2.31 - 1.76 ATSC1e - 3.78 LFEA - 3.42 LFEBH - 0.04 ToPSA$$
$$+ 3.86 MCS48 + 0.47 SubC1 - 0.19 D604 \tag{1.4*}$$

where $\log k_H$ is the logarithm of Henry's law constant values. As shown in eq. (1.4*), Duchowicz et al. [103] used seven complex and unusual descriptors including *ATSC1e, LFEA, LFEBH, ToPSA, MCS48, SubC1*, and *D604*. These descriptors are nonconformational descriptors where they are categorized as follows: (a) *ATSC1e* is a two-dimensional auto-correlation descriptor corresponding to the centered Broto–Moreau autocorrelation-lag 1/weighted by the Sanderson electronegativity; (b) *LFEA* and *LFEBH* are two descriptors of molecular linear free energy relationship where *LFEA* is global hydrogen bonding acidity of the solute, and *LFEBH* is global hydrogen bonding basicity of the solute; (c) *ToPSA* is a topological descriptor corresponding to the topological polar surface area; (d) *MCS48, SubC1*, and *D604* are three indicator descriptors where *MCS48* is the number of OQ(O)O groups in which Q is a heteroatom different from carbon or hydrogen, *SubC1* is the number of primary carbons, and *D604* is the number of substituted sp^2 aromatic carbons. The model of Duchowicz et al. [103] does not require the conformational character-istics information of the molecules for the estimation of Henry's law constant. Equation (1.4*) requires complicated computation procedures for estimating parameters from chemical structures.

In contrast to available QSPR models for different classes of chemicals that require complicated computer codes and descriptors, a simple approach has been introduced for the estimation of Henry's law constant of heterogeneous chemicals including persis-tent pollutants, pesticides, aromatic hydrocarbons, and solvents based on their molecu-lar structures. The new model needs the number of carbon and hydrogen atoms as additive parameters. It also requires the contributions of hydrogen bonding functional groups, polar groups, halogenated compounds, and hydrocarbons as four nonadditive factors. It was given as follows [104]:

$$\log k_H = -2.82 - 0.130\ n_C - 0.760\ n_N - 2.92\ H\ bond - 3.52\ PG + 2.41\ HalC + 2.31\ Hydr$$
$$\tag{1.5*}$$

where n_C and n_N are the number of carbon and hydrogen atoms, respectively; *H bond, PG, HalC*, and *Hydr* are the structural parameters corresponding to hydrogen bonding functional groups, polar groups, halogenated compounds, and hydrocarbon derivatives, respectively. Four nonadditive structural parameters *H bond, PG, HalC*, and *Hydr* are described as follows:

(a) *H bond*: Hydrogen bonding groups including -COOH, -OH, -NH$_2$, and -NH>- can form strong to weak hydrogen bonding with molecules of water. The kind of functional groups and their neighbors has important contributions to $\log k_H$ for the small size of a heterogeneous compound. For large molecular sizes, the strength of hydrogen bonding of heterogeneous compounds can be decreased. Meanwhile, the presence of polar groups or hydrogen bonding groups can enhance interactions with water molecules. Table 1.3 shows different values of *H bond* for chemicals containing hydrogen bonding groups.

Table 1.3: Different values of *H bond* for chemicals containing hydrogen bonding groups.

No.	Group	Structure	Condition	H bond	Example
1	-COOH	R-COOH	R = Hydrocarbon substituent with less than 11 carbon atoms	1.5–0.15 n where n is the number of carbon atoms in R	
			X = Halogen atom	1.6	
		-C(=O)-COOH	–	2.1	
			–	1.4	
2	-OH	R-OH	R = Alkyl group containing more than one -OH or -CHO	1.5	
			R = Alkyl group containing nitrate group	1.0	
			R = Hydrocarbon group containing less than six carbon atoms	0.6	
			R = Hydrocarbon group containing more than five carbon atoms up to nine carbon atoms	0.25	
			R = Alkyl, alkene and alkyne groups containing -O- in the form of ether or -OOH	0.9	

Table 1.3 (continued)

No.	Group	Structure	Condition	H bond	Example
		or	X = Halogen and the alcohol containing less than six carbon atoms	0.5	
		ArCy-OH	ArCy = Aromatic without substituent	1.1	
			ArCy = Aromatic or cyclic hydrocarbon ring with alkyl or halogen substituents	0.7	
			ArCy = Aromatic with alkoxy substituent in the ortho position	0.8	
			ArCy = Aromatic with alkoxy substituent in the non-ortho position	1.2	
			ArCy = Cyclic hydrocarbon ring without substituent	0.5	
			Ar = Aromatic ring	1.6	
			–	1.8	
			–	1.7	
		Ar-(CH$_2$)$_n$-OH	Ar = Aromatic group and $n < 5$	0.9	
3	-NH$_2$	Nirtroaniline or naphthylamine derivatives	Meta or para of -NH$_2$ and -NO$_2$ groups	1.2	
			Naphthylamine, ortho of -NH$_2$ and -NO$_2$ groups as well as naphthylamine derivatives	0.75	

Table 1.3 (continued)

No.	Group	Structure	Condition	H bond	Example
			–	0.95	
			–	0.55	
		X-NH₂	X = A hydrocarbon containing less than four carbon atoms Y = Ar-CO- where Ar is the benzene ring	0.4	-NH₂
		H₂N-X-NH₂ or Y-NH₂		1.45	
4	Aniline, naphthylamine, or amino group attached to cyclic hydrocarbon or alkyl aniline or cycloamine	–	–	0.4	

(b) *PG*: The existence of polar groups such as -S(=O)-, -CO-, -CHO, -Si-O-, -S-F, -CO-NH-, -CN, -P(=O)(O)(S)-, -P(=O)(O)(O)-, -NO₂, -ONO₂, -O-, and -COO- under certain conditions may increase the interaction of a heterogeneous compound with water molecules. For small-size molecules of some heterogeneous compounds such as linear ketone or linear aldehyde, the effects of polar groups on log k_H are important. Table 1.4 provides various values of *H bond* for chemicals including polar groups.

Table 1.4: Different values of *PG*.

No.	Group	Condition	PG	Example
1	-S(=O)- or -CO-CO-	The number of carbon is less than four	1.2	
2	-Si-O-	–	$-0.43\, n_{(Si-O)}$-3.9 where $n_{(Si-O)}$ is the number of Si-O bonds	

Table 1.4 (continued)

No.	Group	Condition	PG	Example
3	or -S-F	–	−0.7	
4	or or or cyclic amide	X = Alkyl group or Y = Cl	0.80	
		X = Aryl group or -SO$_2$- group	1.20	
		Cyclic amide		
5	-CN	The addition of -CN to pyridine ring	0.4	
		The addition of -CN to alkyl group comprising less than three carbon atoms	0.2	
6	or or	R$_1$ and R$_2$ are alkyl groups	1.5	
7	or	Ar = Aromatic ring containing large alkyl group (the number of carbon atoms in alkyl group > 3)	0.45	
8	or	Ar′ = Aromatic ring without substituent or small alkyl groups	−0.4	
9	Nitroaromatic derivatives	Mononitro derivatives without further substituents or alkyl substituents or halogen substituents	0.2	
		More than one nitro group without further substituents or alkyl substituents or halogen substituents	0.65	
10	Pyridine or alkyl pyridine or alkyl pyrazine	–	0.2	
11	or	–	0.7	

Table 1.4 (continued)

No.	Group	Condition	PG	Example
12	or cyclic ether without substituents (or alkyl substituent) or R-O-Ar	R = Alkyl group Ar = Benzene ring without hydrogen bonding group	0.2	
	R_1-X-R_2	R_1 and R_2 are alkyl groups X = O or F	−0.2	
13	Cyclic ketone	The number of carbon atoms is less than seven	0.4	
14	Linear ketone or linear aldehyde	The number of carbon atoms is less than eight	0.25	
15	R-ONO$_2$	R = Alkyl group	−0.4	
16	R-NO$_2$	R = Alkyl group	0.2	
17	The presence of both -COO- and -O- groups or acyclic ketone containing -CF$_3$	–	0.55	

(c) *HalC*: The kind of halogen under certain conditions has essential effects on log k_H. Table 1.5 indicates the optimum values of *HalC* for different types of chemicals.

Table 1.5: The values of *HalC*.

No.	Group	Condition	HalC	Example
1	Fully halogenated alkane	The number of hydrogen atoms is zero	1.7 for fluorine atoms	
			0.5 for chlorine (or bromine) atoms	
			1.3 and 0.9 for the higher contribution of fluorine and chlorine (or bromine) atoms, respectively, as well as 1.1 for an equal number of fluorine and chlorine (or bromine) atoms	

Table 1.5 (continued)

No.	Group	Condition	HalC	Example
2	Polyhalogenated benzene or polyhalogenated naphthalene	Without further substituents	0.3	
3	Monohaloalkanes	–	0.7	
4	F_3C	Containing one carbon atom	0.8 for X = H	
		Containing two carbon atoms	0.8 for X = F	
		Containing three carbon atoms	1.25 for X = F 0.6 for X = Cl	
5	Polyhalogenated biphenyl	1. One ring without a halogen atom or containing one halogen atom 2. The other ring without halogen atom or containing up to four halogen atoms	0.3	
6		–	0.3	
7	Hexahalogenated cyclohexane (different stereoisomers)	–	−0.9	

(d) Hydr: Some classes of hydrocarbon derivatives have higher values of log k_H as compared to those expected from the contribution of n_C, which include three categories:
(1) Saturated and unsaturated nonaromatic hydrocarbons
(2) Benzene, alkyl, and vinyl benzene
(3) Alkyl and vinyl naphthalene

Table 1.6 shows different values of *Hydr* for the three mentioned classes.

Table 1.6: Different values of *Hydr*.

No.	Group	Condition	*Hydr*	Example
1	Benzene, alkyl, and vinyl benzene	–	0.75	
2	Alkyl and vinyl naphthalene	–	0.45	
3	Saturated and unsaturated nonaromatic hydrocarbons	–	1.2	

Example 1.2: 3,5-Dimethylpyridine has the following molecular structure:

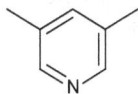

The calculated value of log k_H by eq. **(1.4*)** is –3.43. If the experimental value is –5.19, calculate log k_H using eq. **(1.5*)** and compare the reliability of two equations.

Answer: The values of n_C, n_N, H bond, PG, HalC, and Hydr are 7, 1, 0, 0.2 (Table 1.4, no. 10), 0, and 0, respectively. The use of eq. **(1.5*)** gives:

$$\log k_H = -2.82 - 0.130\, n_C - 0.760\, n_N - 2.92\ H\ bond - 3.52\ PG + 2.41\ HalC + 2.31\ Hydr$$
$$= -2.82 - 0.130(7) - 0.760(1) - 2.92(0) - 3.52(0.2) + 2.41(0) + 2.31(0) = -5.16$$

As shown, the calculated result of eq. **(1.5*)** is closer to the experimental data than **(1.4*)**.

1.8 Common Process for Removal of Organic Compounds in the Environment

Physical, biological, and chemical methods are common three processes to remove some categories of organic compounds such as nitroaromatic compounds in the environment [105]. Physical processes include adsorption, absorption, extraction, photooxidation, ultrafiltration, and volatilization. The adsorption process is widely used as compared to the other processes [106]. Chemical processes contain hydrolysis and advanced oxidation processes (AOPs) where the latter applies oxidants like H_2O_2, Fenton's reagent, and metal as photocatalysts to oxidize organic compounds. It should be mentioned that Fenton's reagent is a solution of H_2O_2 with ferrous iron (such as $FeSO_4$) as a catalyst that is used to oxidize organic compounds as part of AOPs. For AOPs in wastewater treatment, the aromatic ring can be destroyed by radicals generated from oxidants into smaller molecules [107]. The biodegradation process includes aerobic or anaerobic, which uses microorganisms to decompose organic compounds. Some particular organic compounds are tenacious to be oxidized if they are inclined to aerobic degradation because aromatic compounds with high electronegativity can be reduced first. Since the decomposition of organic compounds is solely by bacteria, and microbes differ in their ability to degrade them, the selection of appropriate microorganisms is of vital importance [108]. Adsorbents of physical processes usually adsorb organic compounds altogether. Thus, further treatment of these pollutants is required to detoxify them. Physicochemical and biological processes are integrated to remove organic compounds in the modern industry because AOPs are expensive and microorganisms are inefficient on a high concentration scale [5]. Due to the economic costs, it is essential to assess the toxic effects of organic compounds and to evaluate all the toxic response endpoints through experiments in vitro or in vivo. Since it is impossible to evaluate all the toxic response endpoints experimentally, in silico predictive modeling techniques can reduce the number of animal testing, and satisfy 3R rules (reduction, replacement, and refinement) [109].

1.9 Summary

Section 1.1 described the toxicity measurements of compounds through different types of dose descriptors. Due to the importance of the toxicity of different types of chemicals for environmental hazards, aquatic toxicity was also demonstrated in this section. Section 1.2 demonstrated QSAR/QSPR/QSTR models, which can be applied to predict the biological activity, property, and toxicity of a given set of molecules. Section 1.3 illustrated important statistical parameters for the assessment of QSAR/QSPR/QSTR models through internal and external validations. Since organic solvents are widely used in laboratories and chemical industries, solubility parameters were demonstrated for the selection of nontoxic solvents. Equation (**1.3***) provides a simple approach for the prediction of δ for a desired polymer only from elemental composition and structural parameters of repeating unit structures -(C^1H_2-$C^2R^3R^4$)-. Thus, good solvents containing δ values within ±2 $MPa^{1/2}$ of the calculated polymer δ value can be chosen to increase the concern about safety, effectiveness, and available alternative. Toxicities of PAHs and organophosphate pesticides have been illustrated in Sections 1.5 and 1.6, where hazardous properties of some common derivatives were described. It is valuable to assess accurate knowledge of Henry's law constant because it demonstrates the movement of different types of organic compounds inside and outside aquatic ecosystems. Equation (**1.5***) introduced a simple approach for the estimation of Henry's law constant of heterogeneous chemicals including persistent pollutants, pesticides, aromatic hydrocarbons, and solvents based on their molecular structures. Section 1.8 described physical, biological, and chemical methods, which are common three processes to remove some categories of organic compounds such as nitroaromatic compounds in the environment.

Problems

1. Calculate the solubility parameter δ for the following polymers through eq. (1.3*):

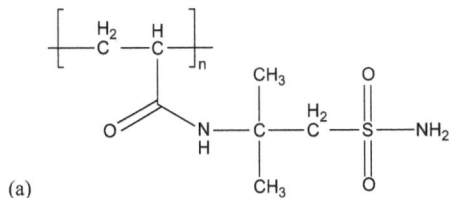

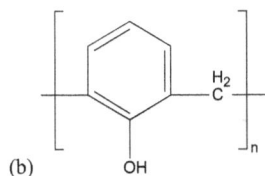

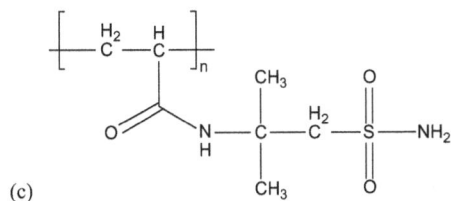

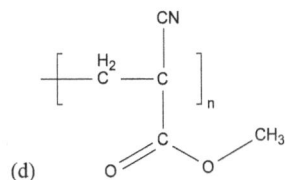

(a)

(b)

(c)

(d)

2. Use eq. (**1.5***) and calculate log k_H for the following compounds:

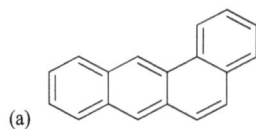

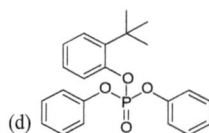

(a)

(b)

(c)

(d)

Chapter 2
Toxicity of Small Data Sets of Organic Compounds

There are some attempts to predict the toxicity of important classes of organic compounds. This chapter reviews important predictive models, which have been derived based on a small data set. These methods are mainly QSAR/QSTR methods, which use the activity/toxicity of a set of compounds as a linear combination of certain descriptors. Each class of organic compounds will be discussed in each section.

2.1 Nitroaromatic Compounds

Nitroaromatic compounds such as TNT (2,4,6-trinitrotoluene) are one of the important classes of energetic compounds, which have wide applications in military and industrial applications [61]. Different methods can be used to estimate the properties of nitroaromatics [110–113]. Nitroaromatics containing one or two nitro groups have wide applications in various industries such as shoe polish, clothing dyes, polymers, solvents for the synthesis of dyes, and plastics or as bioactive products of pesticides, insecticides, and pharmaceuticals because they are insensitive to various external stimuli [114–116]. They may provide some manifestations of toxicity containing immunotoxicity, skin sensitization, and methemoglobinemia [117] or mutagenic [118] and carcinogenic effects, along with less harmful but just as important consequences such as endocrine system impairment, allergic reactions,and skin irritation [119]. Some of their derivatives are also toxic, that is, arylhydroxylamines, arylamines, azo-, or azoxy-compounds [120, 121]. Due to the electron withdrawal of nitro groups, nitroaromatics can resist hydrolysis and chemical or biological oxidation. Thus, they are present everywhere in the environment and cause a potential hazard to human health via pollution of water, air, foods, and sediments [106].

Some efforts have been done to predict the toxicity of nitroaromatics through different models, for example, binary and multiple classification models for the prediction of toxicity nitroaromatics based on LD_{50} values representing acute oral toxicity in experimental rats [122]. There are some QSAR/QSTR methods based on complex descriptors for the prediction of toxicity of nitroaromatics. Kuz'min et al. [123, 124] used Hierarchical Technology for 28 nitroaromatic compounds based on QSAR (HiT QSAR) and 1D QSAR (one-dimensional). Agrawal et al. [125] developed a QSAR model for 40 monosubstituted nitrobenzenes using topological constitutional descriptors. Naizi et al. [126] used ab initio theory for the prediction of toxicity of nitrobenzenes by several quantum-chemical descriptors including electrostatic potentials and local charges at each atom. Since QSAR methods need large numbers of compounds, which are required to extract a good QSAR model [127], Gooch et al. [128] used quantum-chemically calculated descriptors together with molecular descriptors generated by Dragon, PaDEL, and

https://doi.org/10.1515/9783111189673-002

HiT-QSAR software concerning the oral LD_{50} toxicity in rats. Several of the new QSAR/QSTR models for the prediction of toxicity of nitroaromatics are reviewed here.

2.1.1 QSAR/QSTR Studies on Bacteria

2.1.1.1 *Salmonella typhimurium* TA100 Strain

Hao et al. [129] constructed QSAR/QSTR models using a mutagenicity data set of 48 nitroaromatic compounds for the *Salmonella typhimurium TA100 strain*, which was chosen from the Benigni's report (QSARs for mutagenicity and carcinogenicity) for OECD [130]. The Bacterial Ames test without the *S9* activation system determined these mutagenic activities. Hao et al. [129] used Dragon descriptors together with quantum chemistry descriptors to characterize the detailed molecular information. They developed QSAR/QSTR models based on GA and MLR analyses with seven machine-learning methods and six molecular fingerprints. Among different models, they introduced the best QSAR/QSTR model for the prediction of nitroaromatics mutagenicity with five descriptors as follows:

$$\log TA100 = -2.4885 - 38.169E_{HOMO} + 1.2564 Ineffective - 50 + 0.9804 Hpnotic - 80$$

$$+ 0.1974 CATS2D\_04\_LL - 0.0295 TIC2 \tag{2.1*}$$

where E_{HOMO} is the quantum chemistry descriptor; *Ineffective, Hpnotic, CATS2D_04_LL,* and *TIC2* are four Dragon descriptors where the detailed explanations of Dragon descriptors are given in the *Handbook of Molecular Descriptors* [131].

2.1.1.2 *Salmonella typhimurium (TA98)* Bacterial Species with or Without Microsomal Activation (*S9*)

Jillella et al. [132] have compiled two sets of mutagenicity data against *Salmonella typhimurium (TA98)* bacterial species with (*TA98 + S9*) or without (*TA98 − S9*) implementing microsomal activating enzyme named *S9*, solely collected from the literature. They developed QSAR/QSTR models using simple 2D variables having definite physicochemical meaning calculated from Dragon [133, 134], SiRMS, and PaDEL descriptor [133, 135] software tools. Among different models, Jillella et al. [132] used a total of 291 nitroaromatic compounds to derive the following model for *TA98 − S9* in terms of revertants per nanomole (rev nmol^{-1}) containing higher R^2 values:

$$\log (TA98 - S9) \text{ rev nmol}^{-1} = -17.487 + 4.310 RDCHI + 0.326 ETA\_dBeta - 0.040 nCIR$$

$$+ 24.010 X5Av - 0.206 F07[C - C] - 1.892 NRS$$

$$+ 1.390 F09[N - N] - 2.220 B08[Cl - Cl]$$

$$\tag{2.2*}$$

where *RDCHI* and *ETA_dBeta* are descriptors depicting unsaturation in molecules; *nCIR* and *NRS* are descriptors designating the presence of various rings; *X5Av* and *F07[C – C]* are descriptors with more hydrophobic influence; *F09[N – N]* and *B08[Cl – Cl]* are descriptors with more electronegative element contents. Jillella et al. [132] used 309 data of nitroaromatic compounds to derive the following model for *TA98 + S9* containing higher R^2 values:

$$\log(TA98 + S9) \text{ rev nmol}^{-1} = -54.056 + 120.75 ETA\_Epsilon\_3 + 0.224 SaaaC$$

$$+ 0.614 B06[C – C] + 0.010 D/Dtr09 – 1.473(C – 034)$$

$$+ 1.585 F02[N – N] – 0.743 SsssCH – 1.658 nPyridines \quad (2.3^*)$$

where *ETA_Epsilon_3*, *F02[N–N]*, and *nPyridine* are descriptors showing the presence of more electronegative elements; *SaaaC*, *B06[C–C]*, *C-034*, and *D/Dtr09* are descriptors representing hydrophobic moieties; *SsssCH* is a descriptor influencing branching in a molecule.

2.1.2 QSAR/QSTR Studies on Rodents

Lots of QSAR/QSTR models have been constructed by using rats or mice as experimental objects for rodent acute toxicity [122, 124, 129, 136–143]. A simple approach has been introduced to assess the toxicity of nitroaromatic compounds in terms of an oral LD_{50} dose (50% lethal dose) for rats [141]. In contrast to available QSAR/QSTR models for the prediction of *in vivo* toxicity of nitroaromatics where they need complex descriptors, the novel model is based on constitutional descriptors. Experimental data of 90 nitroaromatics are used to derive and test the new model as follows:

$$-\log LD_{50}(oral \; rat, \; mg \; kg^{-1}) = 1.599 + 0.4293 n_{NO_2} – 0.4165 n_S + 1.771 n_P$$

$$+ 1.313 Tox^+ –2.110 Tox^- \quad (2.4^*)$$

where n_{NO_2}, n_S, and n_P are the number of nitro groups, sulfur, and phosphorous atoms in a desired nitroaromatic compound, respectively. Two variables Tox^+ and Tox^- show the contributions of some molecular moieties that can increase and decrease the toxicity of a nitroaromatic compound, respectively. Table 2.1 gives the optimum values of Tox^+ and Tox^-.

Table 2.1: The optimum values of Tox^+ and Tox^-.

Structural parameters	Tox^+	Tox^-
Ar–F	0.8	0
(structures) (–R or –OAr and the specified polar groups in ortho position), (R₁ and R₂ are only alkyl groups)	1.0	0
(structure)	2.0	0.0
(structures) (Y=alkyl group, nitro group, or), (X or Y=nitro group),	0	0.5

Example 2.1: Paraoxon has the following molecular structure:

If the measured value of $-\log LD_{50}(oral\ rat,\ mg\ kg^{-1})$ is 5.184, use eq. (**2.4***) along with Table 2.1 and compare the calculated $-\log LD_{50}(oral\ rat,\ mg\ kg^{-1})$ with the experimental value.

Answer: The use of eq. (**2.4***) gives

$$-\log LD_{50}(oral\ rat,\ mg\ kg^{-1}) = 1.599 + 0.4293n_{NO_2} - 0.4165n_S + 1.771n_P + 1.313Tox^+ - 2.110Tox^-$$

$$= 1.599 + 0.4293(1) - 0.4165(0) + 1.771(1) + 1.313(1.0) - 2.110(0)$$

$$= 5.112$$

As shown, the calculated result of eq. (**2.4***) is close to the experimental data, that is, 5.184.

2.2 Aromatic Aldehydes

Aromatic aldehydes have widespread use causing cytotoxicity, tissue damage, carcinogenicity, and mutagenicity leading to various health complications [144]. They are important intermediates to produce a variety of industrial processes in the pharmaceuticals and agrochemical industries, which are key in the flavor and fragrance industry [145].

Due to their ability to interact with electron-rich biological macromolecules such as nucleic acids and proteins, they have the potential to cause some adverse effects. Since they may have a high potential for environmental pollution, some QSAR/QSTR models have been developed to predict the toxicity of aldehydes against different ecologically relevant organisms [146–151].

2.2.1 Two Descriptors log K_{ow} and the Maximum Acceptor Superdelocalizability in a Molecule

Netzeva and Schultz [146] developed QSAR/QSTR models for acute toxicity to the ciliate *Tetrahymena pyriformis* using mechanistically interpretable descriptors. The resulting QSARs revealed that the 1-octanol/water partition coefficient (log K_{ow}) and electronic descriptor A_{max} (the maximum acceptor superdelocalizability in a molecule) based on calculations with the semiempirical AM1 model are responsible for the demonstration of acute aquatic toxicity of aromatic aldehydes. Since the descriptor log K_{ow} shows hydrophobicity, Netzeva and Schultz [146] used the preferred experimental values or software estimated (ClogP ver. 1.0 for Windows, 1995). Netzeva and Schultz [146] obtained A_{max} in MOPAC 93 (J.J.P. Stewart, Fujitsu Ltd., 1993; Windows 95/98/NT/2k adaptation and MO indices – J. Kaneti, 1988–1994 MO–QC) using AM1 molecular orbital method. Since the group of 2- and/or 4-hydroxylated aldehydes have enhanced toxicity as compared to the other aldehydes, Netzeva and Schultz [146] proposed transformation to quinone-like structures as the explanation for this enhanced potency. They have used 25 data to model the 2- and/or 4-hydroxylated aldehydes as follows:

$$-\log IC_{50}(mM) = -3.11 + 0.540 \log K_{OW} + 8.30 A_{max} \tag{2.5*}$$

where IC_{50} is the 50% growth inhibitory concentration and A_{max} in eV^{-1}. Netzeva and Schultz [146] used the remaining 55 data to model aromatic aldehydes without two- and/or four-position hydroxylated derivatives as follows:

$$-\log IC_{50}(mM) = -4.04 + 0.583 \log K_{OW} + 9.80 A_{max} \tag{2.6*}$$

Example 2.2: (a) Calculate the toxicity of 2-hydroxy-1-naphthaldehyde in terms of $-\log IC_{50}(mM)$. Assume log $K_{ow} = 2.99$ and $A_{max} = 0.3244$ [146]; (b) calculate the toxicity of 5-hydroxy-2-nitrobenzaldehyde in terms of $-\log IC_{50}(mM)$. Assume log $K_{ow} = 1.75$ and $A_{max} = 0.3362$ [146].

Answer: (a) 2-Hydroxy-1-naphthaldehyde has the following molecular structure:

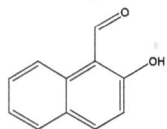

Due to the existence of a hydroxyl group in position 2, eq. (**2.5***) should be used as

$$-\log IC_{50}(mM) = -3.11 + 0.540 \log K_{OW} + 8.30 A_{max}$$
$$= -3.11 + 0.540(2.99) + 8.30(0.3244) = 1.197$$

Answer: (b) Molecular structure of 5-hydroxy-2-nitrobenzaldehyde is given as follows:

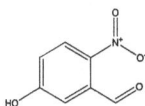

Equation (**2.6***) should be used because there is no –OH group in 2- and/or 4-position of 5-hydroxy-2-nitrobenzaldehyde:

$$-\log IC_{50}(mM) = -4.04 + 0.583 \log K_{OW} + 9.80 A_{max}$$
$$= -4.04 + 0.583(1.75) + 9.80(0.3362) = 0.275$$

2.2.2 Log K_{ow} and Molecular Connectivity Index

Louis and Agrawal [151] established the new linear and nonlinear QSAR/QSTR models to predict the toxicity of aromatic aldehydes in *T. pyriformis*. They have chosen three descriptors by using a stepwise regression method. They found that hydrophobicity, branching of the molecule, and a hydroxyl group at position 2- and/or at position 4 are important factors to influence the toxicity of aromatic aldehydes. They introduced the following MLR model based on log K_{ow} as well as the addition of $\chi1A$ and Ip to improve the statistical quality of the models as follows:

$$-\log IC_{50}(mM) = 4.915 + 0.488 \log K_{OW} - 11.705\chi1A + 0.279Ip \qquad (2.7^*)$$

where $\chi1A$ is the selected connectivity descriptor, which is inversely related to molecular branching and complexity; Ip is indicator parameter where its value is 1 for the presence of 2- and/or 4-hydroxylated aldehyde, otherwise zero.

Example 2.3: Compare the predicted results of eq. (**2.7***) and those obtained in Example 2.2 with experimental data, that is, 1.32 and 0.329 [146] for 2-hydroxy-1-naphthaldehyde, and 5-hydroxy-2-nitrobenzaldehyde, respectively. The values of χ1A are 0.452 and 0.471 for 2-hydroxy-1-naphthaldehyde, and 5-hydroxy-2-nitrobenzaldehyde, respectively. The values log K_{ow} for both compounds are given in Example 2.2.

Answer: The values of Ip for 2-hydroxy-1-naphthaldehyde, and 5-hydroxy-2-nitrobenzaldehyde are 1 and 0 because there is –OH group in 2-hydroxy-1-naphthaldehyde. The use of the values of descriptors in eq. (**2.7***) gives

(i)   2-Hydroxy-1-naphthaldehyde

$$-\log IC_{50}(mM) = 4.915 + 0.488 \log K_{OW} - 11.705\chi 1A + 0.279Ip$$
$$= 4.915 + 0.488(2.99) - 11.705(0.452) + 0.279(1) = 1.362$$

Deviation from experimental data = 1.362 – 1.32 = 0.04

(ii)   5-Hydroxy-2-nitrobenzaldehyde

$$-\log IC_{50}(mM) = 4.915 + 0.488 \log K_{OW} - 11.705\chi 1A + 0.279Ip$$
$$= 4.915 + 0.488(1.75) - 11.705(0.471) + 0.279(0) = 0.256$$

Deviation from experimental data = 0.256 – 0.329 = –0.073

Deviations of the predicted $-\log IGC_{50}(/mM)$ for 2-hydroxy-1-naphthaldehyde, and 5-hydroxy-2-nitrobenzaldehyde from Example 2.2 are –0.123 and –0.054. Thus, eq. (**2.7***) gives a closer prediction for 2-hydroxy-1-naphthaldehyde as compared to experimental data.

2.2.3 Log K_{ow} as well as Electronic and Topological Descriptors

Ousaa et al. [150] used MLR and multiple nonlinear regression (MNLR) for predicting the toxicity of aromatic aldehydes to *T. pyriformis*. The MNLR results provide a better predictive capability. Meanwhile, the MLR results give the most important interpretable results. Among the introduced models, Ousaa et al. [150] introduced the stepwise MLR with two variables as follows:

$$-\log IC_{50}(mM) = -1.928 + 0.709 \log K_{OW} + 0.024y \tag{2.8*}$$

where y is the surface tension in dyn cm^{-1}, which can be calculated by ChemSketch program [152]. As shown in eq. (**2.8***), increasing $\log K_{OW}$ and y are responsible for the greater activity of aromatic aldehydes, presence of electronegative substituents (like O, N, F, Br, and Cl), lipophilic substituents, for example, chlorine. The aldehydic oxygen was also important for toxicity.

Example 2.4: Use eq. (**2.8***) and compare the predicted result with experimental data (the measured $-\log IC_{50}(mM)$ = 1.107) for 5-bromosalicylaldehyde. Assume log K_{ow} and y are 1.842 and 56.90, respectively.

Answer: Equation (**2.8***) gives

$$-\log IC_{50}(mM) = -1.928 + 0.709 \log K_{OW} + 0.024y$$
$$= -1.928 + 0.709(1.842) + 0.024(56.90) = 0.743$$

The deviation of the calculated result from the experimental data is –0.362.

2.3 Amino Compounds

Amino compounds have wide applications in many industries such as agriculture, chemical, food, pharmaceutical, dyestuff, and petrochemical. They contain amino aliphatics and amino aromatics (aniline derivatives), which are used as precursor materials for a wide range of consumer products [153]. Some amino derivatives are highly toxic to living organisms in aquatic systems and soil as well as highly toxic to mammals. Moreover, some of them not only have toxic effects on humans but also can act as carcinogenic and mutagenic [154].

Due to limited data on amino compounds, some attempts have been made to use QSAR/QSTR methods to predict their toxicities. Jackel and Klein [155] predicted oral toxicity data of aliphatic amines and anilines for rats with LD_{50}-values to log K_{ow}, fragment values, and electronic properties. Lu et al. [156] used the quantum-chemical parameters and the topological indices for the prediction of the toxicity of aminobenzenes. Mahani et al. [157] related the oral acute in vivo toxicity of 32 amines and amide drugs to their structural-dependent properties. Lu and Jiaan [158] used the shape, quantum-chemical, and topological descriptors for the assessment of the toxic activities of the aminobenzenes and phenols.

The toxicity of 170 amino compounds in rats via oral LD_{50} has been used to derive a reliable correlation based on some molecular structure parameters as follows [159]:

$$-\log LD_{50}(oral\ rat,\ mg\ kg^{-1}) = -2.801 + 0.464ISTP - 0.467DSTP \qquad (2.9^*)$$

where LD_{50} shows the toxicity of amino compounds in rats via oral (acute toxicity, half lethal dose), which is given as the mass of substance administered per unit mass of test subject (milligrams of substance per kilogram of body mass); the parameters *ISTP* and *DSTP* show increasing and decreasing structural toxicity parameters. The existence of some groups in ortho-, meta-, and para-positions of the amino group is important for aminobenzenes.

2.3.1 Estimation of *ISTP*

(a) Monosubstituted and disubstituted aminobenzenes: *ISTP* = 0.3 for the presence of one substituent (except amino- or chloro-substituted) in meta- and para-positions concerning amino group but *ISTP* = 0.7 for aniline and amino- or chlorosubstituted aniline. *ISTP* = 0.4 for the existence of further substituent in nitro-aminobenzene.

(b) The presence of molecular fragment $CH_3CH_2CH(N)CH_3$: *ISTP* = 1.3.

(c) Primary, secondary, and tertiary amines with general formula R_1-NH_2, R_1-NH-R_2, and $(CH_3)_2\ N-R_1$: *ISTP* = 0.5 if R_1 or R_2 contains more than or equal four carbons.

(d) Cyclic tertiary or secondary amine attached to two unsaturated carbon: *ISTP* = 0.8.

2.3.2 Prediction of *DSTP*

(a) Amino benzenes containing alkyl groups with more than one carbon in which the alkyl group is in the ortho-position to $-NH_2$ group and *ortho*-mononitroaniline: *DSTP* = 0.7.

(b) The existence of more than one $-NH-$ group except for those compounds in which both $-NH-$ groups are attached to secondary carbon: *DSTP* = 1.1.

(c) The presence of $-N-CO-$ and $-N-SO_2-$: The values of *DSTP* are 0.8 and 1.3 for the existence of $-N-CO-$ and $-N-SO_2-$, respectively.

(d) The existence of a triazine ring or three tertiary amines or simultaneously at least one tertiary amine in the form $(CH_3)(R)-N-$ or $(C_2H_5)(R)N-$ with $-NH_2$ or $-N = N-$ group: *DSTP* = 1.0.

(e) The influence $RO-$: *DSTP* = 0.7 for the presence of $RO-$ without the presence of $-COO-$ or $-OH$ groups.

2.3.3 Different Effects of $-OH$ and $-N = O$

(a) The $-OH$ group: (i) *DSTP* = 1.0 for the existence of simultaneously $(CH_3)_2$ N$-$ or $(CH_3)(R)N-$ molecular fragments and $-OH$ group attached to nonaromatic carbon; (ii) *DSTP* = 0.6 for the other situations. *ISTP* = 0.6 for $-OH$ groups attached simultaneously to both aromatic and nonaromatic carbons.

(b) The $-N = O$ group: *DSTP* and *ISTP* are 0.7 and 1.0 for the presence of $-N = O$ in aromatic and nonaromatic compounds, respectively.

Example 2.4: Calculate $-\log LD_{50}(oral\ rat,\ mg\ kg^{-1})$ in rats via oral for the following amino compound:

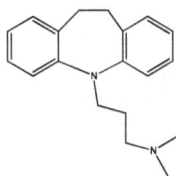

Answer: The use of condition (I) part (d) in eq. (**2.9***) gives

$$-\log LD_{50}(oral\ rat,\ mg\ kg^{-1}) = -2.801 + 0.484ISTP - 0.494DSTP$$
$$= -2.801 + 0.484(0.8) - 0.494(0) = -2.414$$

2.4 Halogenated Phenols

Phenolic compounds are widely used for industrial production as chemical synthesis. Due to their high toxicity and poor biochemical degradability, they have strong teratogenicity, carcinogenicity, and mutagenicity [1–2]. Since phenolic compounds can cause great damage to the environment, animals, plants, and human health, it is necessary to assess and control the toxicity of phenolic compounds. Some efforts have been done to assess the toxicity of halogenated phenols, which are reviewed by three models here.

2.4.1 The DFT-B3LYP Method with the Basis Set 6-31G (d, p), and log K_{OW}

He et al. [160] used the DFT-B3LYP method, with the basis set 6-31G (d, p), to calculate some quantum-chemical descriptors of 43 halogenated phenol compounds. They introduced a suitable correlation for the assessment of the toxicity of these compounds to *T. pyriformis* based on the quantum-chemical descriptors and log K_{OW} as follows:

$$-\log IC_{50}(mM) = -8.893 + 0.75 \log K_{OW} - 30.153 E_{HOMO} + 0.17\mu - 0.538 Q_x \qquad (2.10^*)$$

where IC_{50} is the millimolar concentration causing 50% inhibition of growth about halogenated phenols to *T. pyriformis*; E_{HOMO} is the highest occupied molecular orbital; μ is the dipole moment; Q_x is the total charge of halogens in a molecule. If a halogenated phenol has a high value of log K_{OW}, it gives good lipid solubility. Thus, it penetrates easily into the cell membrane and concentrates on organisms, which increases its toxicity. Since E_{HOMO} measures the donating electron ability, the low value of E_{HOMO} shows the fact that the electrophilic reaction occurs more easily and the toxicity of the halogenated phenol is high. A small value of Q_x provides the bigger toxicity of the halogenated phenol because it has a negative coefficient. Since halogens are withdrawing groups, they draw the π-electron cloud from the benzene ring. Thus, the greater the electrophilicity, the bigger the toxicity of the halogenated phenols. A higher value of the μ can increase the polarity and hydrophilicity of the halogenated phenols. Due to the positive correlative property between μ and the toxicity, the effect between halogenated phenols and organics such as enzymes or proteins is bigger than that between halogenated phenols and water.

2.4.2 Two-Dimensional (2D) and Two Three-Dimensional (3D) QSAR/QSTR Models

Three-dimensional QSAR/QSTR or 3D-QSAR/QSTR models use force field calculations to compute spatial properties of 3D compounds. They provide valuable information on the forces and interactions of molecules [161]. 3D-QSAR/QSTR models can help in designing the new beneficial compounds to screen a large number of chemicals for toxic effects as well as a deeper understanding of the toxicity mechanism. Chen et al. [162] used 2D-QSAR/QSTR and 3D-QSAR/QSTR models using CODESSA program (v2.63, Semichem Inc.) and SYBYL software (v6.9, Tripos Inc.), respectively, to predict the toxicity of halogenated phenols. The results of the 2D-QSAR/QSTR models were consistent with 3D-QSAR/QSTR models. A large number of parameters are used in 2D-QSAR/QSTR models to calculate the desired physiochemical properties from molecules owing to many mature methods including quantum chemistry and molecular mechanism calculation. They are limited because of the lack of descriptions of molecular stereostructure and properties. The 3D descriptors can reflect the situation of molecules more truly. Chen et al. [162] used experimental data of the 50 compounds of halogenated phenols, 50% inhibitory growth concentration (mM) of the ciliate protozoan *T. pyriformis*, to derive and their 2D-QSAR/QSTR and 3D-QSAR/QSTR models. They selected the most important four parameters related to toxicity using stepwise regression based on the training set, finally resulting in the following linear 2D-QSAR/QSTR model:

$$-\log IC_{50}(mM) = -0.653RNH + 0.245MASEH + 0.272MREHO + 1.263\log K_{OW} \quad (2.11^*)$$

where *RNH* shows the relative number of H atoms; *MASEH* indicates the maximum atomic state energy for an H atom; *MREHO* gives maximum resonance energy for an H–O bond. The descriptor *RNH* is directly associated with H bond. *MASEH* corresponds to the maximum value of the atomic valence state energy for the H atom in the molecules. It illustrates the magnitude of the perturbation experienced by an atom in the molecular micro-environment as compared to the isolated atom. *MREHO* relates to the functionality of hydrogen bonding. The four descriptors reflect two important aspects of halogenated phenols: the higher hydrophobicity implies higher toxicity, and the activity of hydrogen in the H–O bond is propitious to decrease toxicity.

Chen et al. [162] also used the 3D-QSAR/QSTR models including comparative molecular field analysis (CoMFA) and comparative molecular similarity analysis (CoMSIA). CoMFA and CoMSIA are two powerful methods in the 3D-QSAR approach. The CoMFA method involves the generation of a common three-dimensional lattice around a set of molecules, which calculates the steric and electrostatic interaction energies at the lattice points. Meanwhile, the CoMSIA method uses the similarity functions represented by Gaussian [163]. They used SYBYL software to calculate the descriptors in 3D-QSAR modeling by a 3D cubic lattice with a regularly spaced grid of 1.0–4.0 °Å units beyond the aligned molecules in all directions. They also calculated the van der Waals potentials and Coulombic terms by using the Tripos force field. For the CoMFA method, an sp³ carbon atom with a van der Walls radius of 1.52 °Å and a charge of 1 e⁻ was a probe atom to calculate steric and electrostatic fields. Chen et al. [162] used the CoMSIA method to calculate the five different fields (steric, hydrophobic, electrostatic, hydrogen-bond donor, and hydrogen-bond acceptor) by using a probe atom of 1 °Å radius with the hydrophobic, charge, and hydrogen-bond properties of +1. The default value of the attenuation factor was 0.3.

2.4.3 Wastewater-Derived Halogenated Phenolic Disinfection By-Products

Due to the increasing number of wastewater-derived aliphatic and phenolic disinfection by-products (DBPs), which were discharged into the aquatic environment with the discharge of disinfected wastewater, Wang et al. [164] investigated the acute toxicity of seven phenolic DBPs from the typical five groups of phenolic DBPs including 2,4,6-trihalo-phenols, 2,6-dihalo-4-nitrophenols, 3,5-dihalo-4-hydroxybenzoic acids, 3,5-dihalo-4-hydroxybenzaldehydes, and halo-salicylic acids as well as four aliphatic DBPs to *Gobiocypris rarus*. They have found two linear correlations between –log LC_{50}, the 96 h experimental values of the half lethal concentration (LC_{50}), and log D_{OW} where log D_{OW} is the logarithm of the *n*-octanol/water distribution coefficient and calculated by the Calculator Plugins from MarvinSketch 15.6.29.0, 2015, ChemAxon (http://www.chemaxon.com) at pH = 7.50, as follows:

$$-\log LC_{50}(\textit{phenolic and aliphatic DBPs, mg } L^{-1}) = -1.62 + 0.42 \log D_{OW} \qquad (R^2 = 0.746)$$

$$(\textbf{2.12}^*)$$

$$-\log LC_{50}(\textit{phenolic DBPs, mg } L^{-1}) = -2.01 + 0.602 \log D_{OW} \qquad (R^2 = 0.966)$$

$$(\textbf{2.13}^*)$$

Example 2.5: (a) Use eqs. (**2.12***) and (**2.13***) and calculate –log $LC_{50}(mg\ L^{-1})$ to *Gobiocypris rarus* for 2,4,6-trichlorophenol; (b) if the experimental value of LC_{50} is 3.70 mg L^{-1}, which equation gives a more reliable prediction? Assume log D_{OW} = 2.06 [164].

Answer: (a) Equation (**2.12***) gives

$$-\log LC_{50}(\textit{phenolic and aliphatic DBPs, mg } L^{-1}) = -1.62 + 0.42 \log D_{OW}$$

$$= -1.62 + 0.42(2.06) = -0.75$$

$$LC_{50}(\textit{phenolic and aliphatic DBPs, mg } L^{-1}) = 5.62$$

Equation (**2.13***) provides

$$-\log LC_{50}(\textit{phenolic DBPs, mg } L^{-1}) = -2.01 + 0.602 \log D_{OW}$$

$$= -2.01 + 0.602(2.06) = -0.77$$

$$LC_{50}(\textit{phenolic DBPs, mg } L^{-1}) = 5.80$$

(b) Equation (**2.12***) gives a closer value to experimental data.

2.5 Organophosphate Compounds

As mentioned in Section 1.6, organophosphate compounds are commonly used as pesticides. Several QSAR/QSTR models have been introduced for the evaluation of the acute toxicity of organophosphate compounds. Model development to predict the toxicity of organophosphate insecticides in different matrices was carried out mainly using the MLR method. Bermúdez-Saldaña and Cronin [165] used MLR and PLS methods to construct the QSAR model of acute toxicity to rainbow trout with 75 organophosphate compounds and carbamates. They found that knowledge of the incorporating mechanism can improve the model's prediction ability. Senior et al. [166] used the MLR approach to employ quantum-chemical and topological descriptors for predicting the toxicity of nine organophosphate compounds. Three molecular field analysis methods have been used through different 3D-QSAR models to assess the acute toxicity of 35 organophosphate compounds in houseflies [167]. Can [168] used QSAR/QSTR models for the estimation of the acute oral toxicity of organophosphate insecticides to male rats. For the development of the models, the 20 chemicals of the training set and the seven compounds of the external testing set were described using descriptors for lipophilicity, polarity, and molecular geometry, as well as quantum-chemical descriptors for energy. Hydrogen bonding between the O atom and NH group in acetylcholinesterase, the major target of organophosphate compounds, as well as electrostatic and steric effects, have significant effects on the reaction between organophosphate compounds and acetylcholinesterase, which are related to the acute toxicity of organophosphate compounds. Chemical reactivity descriptors based on conceptual density-functional theory (DFT) have certain application potential in the acute toxicity prediction of organophosphate compounds [169, 170]. Since the statistical algorithms of the mentioned models are based on relatively small data sets, Wang et al. [138] used the acute toxicity data of 161 organophosphate compounds in two species via six different administration routes to develop a series of QSAR/QSTR models. They developed six QSAR/QSTR models for each route in a single species and two QSAR/QSTR models for a single route in the two species, which achieved practical predictive performance. Kianpour et al. [171] developed the QSAR/QSAR models based on genetic algorithm-multiple linear regression (GA-MLR) and back-propagation artificial neural network (BPANN) methods for the prediction of oral acute toxicity of organophosphates (including some pesticides and insecticides). They found that molecular descriptors obtained by BPANN models could well characterize the molecular structure of each compound.

Substitute species are used in the risk assessment of different types of chemicals because they reduce *in vivo* use of animals in toxicology. They use the results obtained using direct or indirect relationships from different toxicity tests [172]. Interspecies quantitative structure–toxicity–toxicity (QSTTR) modeling can predict toxicity to several other species using the experimental toxicity values of one species [173]. The interspecies QSTTR modeling can reduce the use of higher organisms and understanding of the mechanism of toxic action. Since they extrapolate the data for one toxicity endpoint to those for another toxicity endpoint, they can determine the species-specific toxicity of a compound [174].

Ilia et al. [175] developed interspecies QSTTR models using experimental rat and mouse acute toxicity data of 76 organophosphorus compounds with diverse structures for estimation of the oral acute toxicity to a particular species using available experimental data toward a different species. They extrapolated the known toxicity of chemicals of interest to species missing toxicity data. They used the MLR method to relate the mouse acute toxicity data of organophosphorus compounds with the rat acute toxicity as follows:

$$-\log LD_{50}(oral\ mouse,\ mol\ kg^{-1}) = 0.768 - 0.775\ \log LD_{50}(oral\ rat,\ mol\ kg^{-1}) \quad (2.14^*)$$

where $LD_{50}(oral\ mouse,\ mol\ kg^{-1})$ is the estimated oral mouse acute toxicity (mol kg^{-1}) and $LD_{50}(oral\ rat,\ mol\ kg^{-1})$ is the experimental oral rat acute toxicity (mol kg^{-1}). Ilia et al. [175] improved the reliability of the eq. (2.14*) by considering several complex descriptors. Equations (2.15*) and (2.16*) show two well-developed correlations by considering two and three further descriptors as follows:

$$-\log LD_{50}(oral\ mouse,\ mol\ kg^{-1}) = 0.581 - 0.811\ \log LD_{50}(oral\ rat,\ mol\ kg^{-1})$$
$$- 0.0713 Mor06m + 0.00489 TPSA(NO) \quad (2.15^*)$$

$$-\log LD_{50}(oral\ mouse,\ mol\ kg^{-1}) = 0.629 - 0.809\ \log LD_{50}(oral\ rat,\ mol\ kg^{-1})$$
$$- 0.0645 Mor06m + 0.00474 TPSA(NO)$$
$$+ 0.394 Mor026m \quad (2.16^*)$$

where two Mor06m and Mor26m are two 3D-MorSE descriptors, where Mor06m and Mor26m represent 3D-MoRSE-signal 06/weighted and 3D-MoRSE-signal 26/weighted by atomic masses, respectively; TPSA(NO) is one molecular property, which represents the topological polar surface area using N, O polar contributions. As shown in eqs. (2.15*) and (2.16*), increasing the value of the Mor06m descriptor value would lead to lower acute toxicity. Meanwhile, the increment of TPSA(NO) descriptor value in eqs. (2.15*) and (2.16*) as well as Mor26m in eq. (2.16*) raise the organophosphate compounds toxicity.

Example 2.6: (a) Use eqs. (2.14*), (2.15*), and (2.16*) to calculate $LD_{50}(oral\ mouse,\ mol\ kg^{-1})$ of the following organophosphate compound:

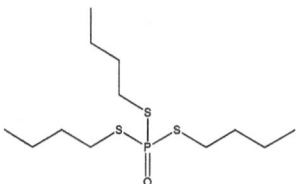

Assume that the values of $-\log LD_{50}(oral\ rat,\ mol\ kg^{-1})$, Mor06m, TPSA(NO), and Mor26m are 3.32, −0.997, 17.07, and 0.328, respectively. (b) If the experimental value of $-\log LD_{50}(oral\ mouse,\ mol\ kg^{-1})$ is 3.61 [175], calculate deviations of eqs. (2.14*), (2.15*), and (2.16*).

Answer: (a) Equation (**2.14***) gives

$$-\log LD_{50}(\text{oral mouse, mol kg}^{-1}) = 0.768 - 0.775 \ \log LD_{50}(\text{oral rat, mol kg}^{-1})$$

$$= 0.768 - 0.775 \ (-3.32) = 3.34$$

Equation (**2.15***) provides

$$-\log LD_{50}(\text{oral mouse, mol kg}^{-1}) = 0.581 - 0.811 \ \log LD_{50}(\text{oral rat, mol kg}^{-1}) - 0.0713 Mor06m$$

$$+ 0.00489 TPSA(NO)$$

$$-\log LD_{50}(\text{oral mouse, mol kg}^{-1}) = 0.581 - 0.811 \ (-3.32) - 0.0713 \ (-0.997)$$

$$+ 0.00489 \ (17.07) = 3.43$$

Equation (**2.16***) offers

$$-\log LD_{50}(\text{oral mouse, mol kg}^{-1}) = 0.629 - 0.809 \log LD_{50}(\text{oral rat, mol kg}^{-1}) - 0.0645 Mor06m$$

$$+ 0.00474 TPSA(NO) + 0.394 Mor026m$$

$$-\log LD_{50}(\text{oral mouse, mol kg}^{-1}) = 0.629 - 0.809 \ (-3.32) - 0.0645 \ (-0.997)$$

$$+ 0.00474 \ (17.07) + 0.394 \ (0.328) = 3.59$$

(b) Equation (**2.14***): Dev = −0.27; eq. (**2.15***): Dev = −0.18; eq. (**2.16***): Dev = −0.02. Thus, eq. (**2.16***) gives a more reliable result.

2.6 Polychlorinated Naphthalenes

Polychlorinated naphthalenes are a subgroup of chlorinated polycyclic aromatic hydrocarbons, which may be an extensive group of little-known environmental pollutants, which are also accumulated in biota. They can form up to 75 analogs containing one to eight chlorine atoms per naphthalene molecule. They are persistent, toxic, and lipophilic [176]. They can be used as dielectric fluids for capacitors and flame retardants, dye-making, fungicides in the wood, and paper industries, textile and plasticizers, casting materials for alloys, and lubricants for graphite electrodes and oil additives [177]. They can be produced as trace contaminants in commercial polychlorinated biphenyl formulations [176]. They can also be formed in combustion products such as fly ash, flue gas, and circulating water of the solid waste incinerator [178].

Nath et al. [179] introduced several 2D-QSAR/QSTR models by employing 75 polychlorinated naphthalenes against green algae, *Daphnia magna*, and fish of the aquatic food web. They derived 2D-QSAR/QSTR models based on half-time effective concentration (EC_{50}) at 96 h for green algae, half-time lethal concentration (LC_{50}) at 48 h for *D. magna*, and LC_{50} at 96 h for fish. They found out that bulkiness or molecular size, average functionality index, frequency of Cl–Cl bond at topological distance 7, electronic features, the third-order average connectivity index, and lipophilicity are the significant features of polychlorinated naphthalenes that affect the aquatic toxicity toward green algae, *D. magna* and fish.

For the development of the models, Nath et al. [179] calculated only 2D descriptors employing E-state indices, atom-centered fragments, connectivity, constitutional, ring, 2D atom pairs, functional, topological, molecular property and extended topochemical atoms (ETA) descriptors employing alvaDesc (v2.0) software (https://www.alvascience.com/alvadesc/) from Alvascience Srl. The best 2D-QSAR/QSTR models employing only non-ETA 2D descriptors are given as follows:

$$-\log EC_{50}(green\ algae,\ M) = -0.68898 + 3.90242X3A + 1.11826MLOGP \qquad (2.17^*)$$

$$-\log LC_{50}(Daphnia,\ M) = -1.16197 + 1.24168X3A + 1.38527MLOGP \qquad (2.18^*)$$

$$-\log LC_{50}(fish,\ M) = -1.5931 + 1.20616X3A + 1.44861MLOGP \qquad (2.19^*)$$

where *X3A* is calculated based on Kier and Hall's connectivity index [180], which defines the "average connectivity index of order 3"; *MLOGP* or log $K_{OW,M}$ is described as the Moriguchi octanol–water partition coefficient. *X3A* is a subtype of connectivity descriptors that encrypts the "chi" value across three bonds. *MLOGP* evaluates the lipophilicity of molecules because the addition of halogen substituents will increase the lipophilicity of the molecules [181]. *MLOGP* can be calculated by counting lipophilic atoms, including carbon with the multiplier rule for their contributions, and hydrophilic atoms containing all nitrogen and oxygen atoms [182]. It measures lipophilicity, which characterizes the penetration ability of the compound through a lipid-rich zone from an aqueous solution. Since *MLOGP* has a positive coefficient, the toxicity of investigated compounds increases with increasing lipophilicity.

Nath et al. [179] investigated the interrelationships among the toxicity endpoints of the aquatic toxicity toward green algae, *D. magna*, and fish through interspecies modeling using 75 common polychlorinated naphthalenes. They developed six different MLR models as follows:

$$-\log EC_{50}(green\ algae,\ M) = 0.85136 - 0.79599 \log LC_{50}(Daphnia,\ M) \qquad (2.20a^*)$$

$$-\log LC_{50}(fish,\ M) = -0.39307 - 1.04546 \log LC_{50}(Daphnia,\ M) \qquad (2.20b^*)$$

$$-\log EC_{50}(Daphnia,\ M) = 0.3736 - 0.95646 \log LC_{50}(fish,\ M) \qquad (2.20c^*)$$

$$-\log EC_{50}(green\ algae,\ M) = 1.15064 + 0.76137 \log LC_{50}(fish,\ M) \qquad (2.20d^*)$$

$$-\log EC_{50}(Daphnia,\ M) = 1.01304 - 1.24701 \log EC_{50}(green\ algea,\ M) \qquad (2.20e^*)$$

$$-\log LC_{50}(fish,\ M) = -1.45257 - 1.30377 \log EC_{50}(green\ algea,\ M) \qquad (2.20f^*)$$

Equations (**2.20a***) to (**2.20f***) are extremely robust and predictive as evidenced by their promising internal as well as external validation metrics.

2.7 Assessment of the Agonistic Activity of Dibenzazepine Derivatives

An agonist is a substance that activates a receptor to yield a biological response. Receptors are cellular proteins whose activation causes the cell to change what it is currently doing. The nonselective cation channel transient receptor potential ankyrin 1 (TRPA1) receptor is a member of the transient receptor potential (TRP) family of cation-selective channels, which were shown to transduce mechanical, thermal, and pain-related inflammatory signals [183]. TRPA1 receptor is primarily stated in small diameter, nociceptive neurons [184–186], that have a key role as a biological sensor because it is implicated in a growing number of pathophysiological conditions [187], for example, airway diseases [188], neuropathic or inflammatory pain [189], and bladder disorders [190]. Activation of TRPA1 may contribute to the perception of noxious stimuli and inflammatory hyperalgesia such as isothiocyanates like allylisothiocyanate present in mustard oil (MO), wasabi, and horseradish [191]; methyl salicylate available from wintergreen oil [192]; Δ^9-tetrahydrocannabinol as the psychoactive compound available in marijuana [193]; cinnamaldehyde present in cinnamon [194]; acrolein as an irritant found in vehicle exhaust fumes and tear gas [195]; allicin and diallyl disulfide present in garlic [196]; the lacrimators including 1-chloroacetophenone (CN), dibenz[b,f]-[1,4]oxazepine (CR) and 2-chlorobenzylidene malononitrile (CS) [197]. Most of the known activators are electrophilic chemicals except some nonelectrophilic agents comprising thymol and menthol as TRPA1 agonists. The TRPA1 receptor can be activated by the formation of a reversible covalent bond with cysteine residues existing in the active site of the TRPA1 channel [198]. Some investigations developed reversible ligands as agonists with target TRPA1 receptors under pharmacological evaluation [199]. A series of compounds comprising 11H-dibenz[b,e]azepines (morphanthridines) and dibenz[b,f][1,4]oxazepines can act as extremely potent activators of the human TRPA1 receptor activities [200]. These compounds provide insight into the structure–activity relationship (SAR) around this class of TRPA1 agonists with half maximal effective concentration (EC_{50}) values ranging from 1 μM to 0.1 nM.

Ai et al. [200] used CoMFA and CoMSIA methods as 3D-QSAR approaches for the prediction of the agonistic activities of these agonists of 11H-dibenz[b,e]azepine and dibenz[b,f][1,4]-oxazepine derivatives. They used CoMFA and CoMSIA to draw contour maps for identifying important regions where changes in the steric and electrostatic fields can increase or decrease the activity. The CoMSIA model depends on complex descriptors steric (S), electrostatic (E), hydrophobic (H), hydrogen bond donor (D), and acceptor (A) fields but the CoMFA model needs the first two descriptors to visualize contour maps. Novak et al. [201] used UV photoelectron spectroscopy and quantum chemistry calculations to study the electronic structures and conformers of several dibenzazepines. They have related the changes in the electronic structure to the observed biological activity. A simple model has been introduced for the prediction of

EC_{50} of 11H-dibenz[b,e]azepine and dibenz[b,f][1,4]-oxazepine derivatives using specific molecular moieties as follows [202]:

$$-\log EC_{50} = 8.487 + 1.051pEC^+_{50,CR} - 0.987pEC^-_{50,CR} \qquad (2.21^*)$$

where $pEC^+_{50,CR}$ and $pEC^-_{50,CR}$ are two structural descriptors. The values of $pEC^+_{50,CR}$ and $pEC^-_{50,CR}$ are given in Tables 2.2 and 2.3. Tables 2.4 and 2.5 show the assessments of $-\log EC_{50}$ through various substituents for different positions of 11H-dibenz[b,e]azepine and dibenz[b,f][1,4]-oxazepine derivatives, respectively. The values of $-\log EC_{50}$ for positions 1, 2, 3, 4, 7, and 8 are related to the other positions.

Table 2.2: The values of $pEC^+_{50,CR}$ and $pEC^-_{50,CR}$ for 11H-dibenz[b,e]azepine derivatives.

		$pEC^+_{50,CR}$	
Substituent	**1**	**2**	**10**
–COOCH$_3$	A (= 0.7)	A + 0.2	1.5
–COOR, –CONHR	B	B + 0.2	–
–Br	C	C + 0.2	–
–CN, –CONH$_2$	D (= 0.5)	D + 0.2	0.2
–CONR$_2$	E (= 0.0)	E + 0.2	–
–COOH	F	F + 0.2	–
–OCH$_3$	G	G + 0.2	–

	$pEC^-_{50,CR}$				
Substituent	**3**	**4**	**7**	**8**	**9**	**10**
–COOCH$_3$	A′ (= 0.8)	A′ + 0.2	A′ + 0.2	A′ + 0.7	0.2	–
–COOR, –CONHR	B′ (= 1.8)	B′ + 0.2	B′ + 0.2	B′ + 0.7	–	0.3
–Br	C′ (= 0.3)	C′ + 0.2	C′ + 0.2	C′ + 0.7	0	0.1
–CN, –CONH$_2$	D′ (= 1.0)	D′ + 0.2	D′ + 0.2	D′ + 0.7	–	–
–CONR$_2$	E′	E′ + 0.2	E′ + 0.2	E′ + 0.7	2.7	1.1
–COOH	F′	F′ + 0.2	F′ + 0.2	F′ + 0.7	–	1.9
–OCH$_3$	G′	G′ + 0.2	G′ + 0.2	G′ + 0.7	–	–
Without substituent		$pEC^+_{50,CR}$=0.0 and $pEC^-_{50,CR} = 0.0$				

Table 2.3: The values of $pEC_{50,CR}^+$ and $pEC_{50,CR}^-$ for dibenz[b,f][1,4]-oxazepine derivatives.

$pEC_{50,CR}^+(X = O)$

Substituent	1	2	10
–COOCH$_3$	a ($A + 0.8 = 1.5$)	$a + 0.2$	1.7
–COOR, –CONHR	b ($B + 0.8$)	$b + 0.2$	1.4
–Br	c ($C + 0.8$)	$c + 0.2$	–
–CN, –CONH$_2$	d ($D + 0.8$)	$d + 0.2$	–
–CONR$_2$	e ($E + 0.8$)	$e + 0.2$	–
–COOH	f ($F + 0.8$)	$f + 0.2$	–
–OCH$_3$	g ($G + 0.8$)	$g + 0.2$	–

$pEC_{50,CR}^-$ $(X = O)$

Substituent	3	4	7	8	9	10
–COOCH$_3$	a'	$a' + 0.2$	$a' + 0.2$	$a' + 0.7$	0.3	–
–COOR, –CONHR	b'	$b' + 0.2$	$b' + 0.2$	$b' + 0.7$	–	–
–Br	c'	$c' + 0.2$	$c' + 0.2$	$c' + 0.7$	–	–
–CN, –CONH$_2$	d'	$d' + 0.2$	$d' + 0.2$	$d' + 0.7$	–	–
–CONR$_2$	e'	$e' + 0.2$	$e' + 0.2$	$e' + 0.7$	–	1.1
–COOH	f'	$f' + 0.2$	$f' + 0.2$	$f' + 0.7$	–	–
–OCH$_3$	g'	$g' + 0.2$	$g' + 0.2$	$g' + 0.7$	0.7	–
Without substituent (X = O)			$pEC_{50,CR}^+$=1.0 and $pEC_{50,CR}^-$ = 0.0			
Without substituent (X = S)			$pEC_{50,CR}^+$=0.0 and $pEC_{50,CR}^-$ = 0.5			

Table 2.4: The values of pEC_{50} for different positions of 11H-dibenz[b,e]azepine derivatives.

-COOCH$_3$ ($pEC_{50,CR}^-$=A'=0.8)

-COOR, -CONHR ($pEC_{50,CR}^-$=B'=1.8)

-Br ($pEC_{50,CR}^-$=C'=0.3)

-CN, -CONH$_2$ ($pEC_{50,CR}^-$=D'=1.0)

-CONR$_2$ ($pEC_{50,CR}^-$=E')

-COOH ($pEC_{50,CR}^-$=F')

pEC_{50}=8.847-0.987[($A'/B'/C'/D'/E'/F'$)+0.2]

pEC_{50}=8.847-0.987[($A'/B'/C'/D'/E'/F'$)+0.2]

pEC_{50}=8.847-0.987[($A'/B'/C'/D'/E'/F'$)]

pEC_{50}=8.847-0.987[($A'/B'/C'/D'/E'/F'$)+0.7]

pEC_{50}=8.847+1.051[($A/B/C/D/E/F$)+0.2]

pEC_{50}=8.847+1.051[($A/B/C/D/E/F$)]

Decreasing pEC_{50}

Decreasing pEC_{50}

Increasing pEC_{50}

-COOCH$_3$ ($pEC_{50,CR}^+$=A=0.7)

-COOR, -CONHR ($pEC_{50,CR}^+$=B)

-Br ($pEC_{50,CR}^+$=C)

-CN, -CONH$_2$ ($pEC_{50,CR}^+$=D=0.5)

-CONR$_2$ ($pEC_{50,CR}^+$=E=0.0)

-COOH ($pEC_{50,CR}^+$=F)

Table 2.5: The values of pEC_{50} for different positions of dibenz[b,f][1,4]-oxazepine derivatives.

-COOCH$_3$ ($pEC_{50,CR}^-$=a')

-COOR, -CONHR ($pEC_{50,CR}^-$=b')

-Br ($pEC_{50,CR}^-$=c')

-CN, -CONH$_2$ ($pEC_{50,CR}^-$=d')

-CONR$_2$ ($pEC_{50,CR}^-$=e')

-COOH ($pEC_{50,CR}^-$=f')

pEC_{50}=8.847-0.987[(a'/b'/c'/d'/e'/f')+0.2]

pEC_{50}=8.847-0.987[(a'/b'/c'/d'/e'/f')+0.2]

pEC_{50}=8.847-0.987[(a'/b'/c'/d'/e'/f')+0.1]

pEC_{50}=8.847-0.987[(a'/b'/c'/d'/e'/f')]

pEC_{50}=8.847+1.051[(a/b/c/d/e/f)+0.2]

pEC_{50}=8.847+1.051[(a/b/c/d/e/f)]

Decreasing pEC_{50}

Decreasing pEC_{50}

Increasing pEC_{50}

-COOCH$_3$ ($pEC_{50,CR}^+$=a=A+0.8=1.5)

-COOR, -CONHR ($pEC_{50,CR}^+$=b=B+0.8)

-Br ($pEC_{50,CR}^+$=c=C+0.8)

-CN, -CONH$_2$ ($pEC_{50,CR}^+$=d=D+0.8=1.3)

-CONR$_2$ ($pEC_{50,CR}^+$=e=E+0.8=1.5)

-COOH ($pEC_{50,CR}^+$=f=F+0.8)

Example 2.6: (a) Use eq. (**2.21***) for calculation of pEC_{50} of the following compound:

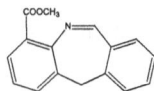

(b) If the calculated values of pEC_{50} by two methods CoMFA [200] and CoMSIA [200] were 7.587 and 7.526, respectively, calculate the deviations of three models as compared to experimental data, that is, 7.5086 [203].

Answer: (a) Equation (**2.21***) using Table 2.2 or Table 2.4 gives

$$-\log EC_{50} = 8.487 + 1.051pEC_{50,CR}^+ - 0.987pEC_{50,CR}^-$$

$$-\log EC_{50} = 8.487 + 1.051(0) - 0.987(0.8 + 0.2) = 7.500$$

(b) Equation (**2.21***): dev = 0.009

CoMFA: dev = −0.078

CoMSIA: dev = −0.017

2.8 Summary

Several QSAR/QSTR methods have been reviewed in this chapter for the assessment of the toxicity of small data sets of organic compounds. Section 2.1 demonstrated different approaches for the prediction of toxicity of nitroaromatic compounds, which include QSAR/QSTR models for bacteria, and rodents. Equation (2.4*) introduced a simple approach to predict the toxicity of nitroaromatic compounds through oral LD_{50} dose for rats from five descriptors n_{NO_2}, n_S, n_P, Tox^+, and Tox^-. Equations (2.5*), (2.6*), (2.7*), and (2.8*) were four QSAR/QSTR models for the prediction of toxicity of aromatic aldehydes to the ciliate *T. pyriformis* using $\log K_{OW}$ and diverse descriptors including A_{max}, γ, $\chi 1A$, and Ip. Equation (2.9*) introduced two parameters $ISTP$ and $DSTP$ for the assessment of toxicity of amino compounds based on LD_{50} in rats via oral. Toxicity assessments of halogenated phenols were done by several paths through the ciliate protozoan *T. pyriformis*, which include (i) the DFT-B3LYP method, with the basis set 6-31G (d, p), and $\log K_{OW}$; (ii) 2D-QSAR/QSTR and 3D-QSAR/QSTR models. Equations (2.12*) and (2.13*) provided two linear relationships between LC_{50} and $\log D_{OW}$ of DBPs to *Gobiocypris rarus*. The three equations (2.14*), (2.15*), and (2.16*) were the three interspecies QSTTR models for the valuation of toxicity of organophosphorus compounds, which relate LD_{50}(oral mouse, mol kg^{-1}) to LD_{50}(oral rat, mol kg^{-1}). Section 2.6 introduces three 2D-QSAR/QSTR models for the assessment of toxicity of polychlorinated naphthalenes against green algae (EC_{50}), *D. magna* (LC_{50}), and fish of the aquatic food web (LC_{50}). Moreover, five simple interspecies QSTTR models including (2.20a*) to (2.20f*) of polychlorinated naphthalenes were introduced in this section. Assessment of the agonistic activity of dibenzazepine derivatives was done through structural descriptors $pEC^+_{50,CR}$ and $pEC^-_{50,CR}$ in eq. (2.21*).

Problems

1. (a) Calculate an oral LD_{50} dose *(50% lethal dose)* for rats of Parathion methyl with the following molecular structure using (2.4*):

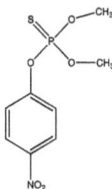

(b) If the measured and the calculated $-\log LD_{50}$*(oral rat, mg kg^{-1})* by [128] were 4.641 and 3.712, which model gives a more reliable result?

2. (a) Calculate the toxicity of 3-ethoxy-4-hydroxybenzaldehyde in terms of $-\log IC_{50}(mM)$. Assume $\log K_{ow} = 1.58$ and $A_{max} = 0.3170$. (b) Calculate the toxicity of 3-hydroxy-4-methoxybenzaldehyde in terms of $-\log IC_{50}(mM)$. Assume $\log K_{ow} = 0.97$ and $A_{max} = 0.3169$.

3. Compare the predicted results of eq. **(2.7*)** and those obtained in Problem 2 with experimental data, that is, 0.015 and −0.142 [146] for 3-ethoxy-4-hydroxybenzaldehyde, and 3-hydroxy-4-methoxybenzaldehyde, respectively. The values of $\chi 1A$ are 0.481 and 0.479 for 3-ethoxy-4-hydroxybenzaldehyde, and 3-hydroxy-4-methoxybenzaldehyde, respectively. The values $\log K_{ow}$ for both compounds are given in Problem 2.

4. Use eq. **(2.8*)** and compare the predicted result with experimental data (the measured $-\log IC_{50}(mM)=0.617$) for 5-bromovanillin. Assume that $\log K_{ow}$ and y are 1.973 and 51.60, respectively.

5. Calculate $-\log LD_{50}(mg\ kg^{-1})$ in rat via oral for the following amino compound:

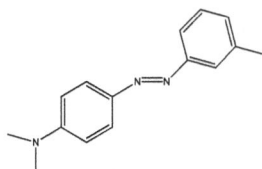

6. (a) Use eqs. **(2.12*)** and **(2.13*)** and calculate $-\log LC_{50}(mg\ L^{-1})$ to *Gobiocypris rarus* for 2,4,6-tribromophenol. (b) If the experimental value of LC_{50} is 1.43 mg L^{-1}, which equation gives a more reliable prediction? Assume $\log D_{ow} = 2.83$ [164].

7. (a) Use eqs. **(2.14*)**, **(2.15*)**, and **(2.16*)** to calculate $LD_{50}(oral\ mouse,\ mol\ kg^{-1})$ of the following organophosphate compound:

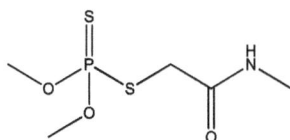

Assume that the values of $-\log LD_{50}(oral\ rat,\ mol\ kg^{-1})$, *Mor06m*, *TPSA(NO)*, and *Mor26m* are 3.06, 0.159, 47.56, and −0.115, respectively. (b) If the experimental value of $-\log LD_{50}(oral\ mouse,\ mol\ kg^{-1})$ is 3.58 [175], calculate the deviations of **(2.14*)**, **(2.15*)**, and **(2.16*)**.

8. (a) Use eq. **(2.21*)** for the calculation of pEC_{50} of the following compound:

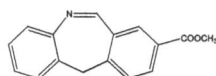

(b) If the calculated values of pEC_{50} by two methods CoMFA [200] and CoMSIA [200] were 6.982 and 7.028, respectively, calculate the deviations of three models as compared to experimental data, that is, 7.5086 [203].

Chapter 3
Toxicity of Medium-Sized Data Sets

Since there are much experimental data for the toxicity of some classes of organic compounds with medium-sized data sets, this chapter reviews important predictive models based on quantitative structure–activity/property/toxicity relationships (QSAR/QSTR) models. These models will be discussed in several sections.

3.1 Polycyclic Aromatic Hydrocarbons (PAHs)

The toxicities of polycyclic aromatic hydrocarbons (PAHs) were described in Section 1.5. PAHs may be produced from incomplete combustion or thermal decomposition under the reductive condition of organic compounds and fossil fuels. The other sources of PAHs are cigarette smoke, automobile emission, and the secondary reaction of parent PAHs with NO_x or OH radicals as well as the use or treatment of pesticides, pharmaceuticals, coatings, and agricultural fertilizers [204]. Due to the high molecular weight and low vapor pressure of PAHs, they are nonvolatile and resistant to chemical reactions and biodegradation [205].

Several QSAR/QSTR studies were done to find the correlations between the toxicity of PAHs and complex descriptors but they used mainly small data sets [206, 207]. Sun et al. [208] developed several QSAR/QSTR models by genetic algorithm (GA) and multiple linear regression (MLR) for the rat's acute oral toxicity (LD_{50}) of PAHs. For deriving and testing the best QSAR/QSTR, they used 106 training set compounds and 20 test set compounds. They introduced the following correlation as the best QSAR/QSTR model comprising eight 2D descriptors, which show maximum atom-type van der Waals surface area, electrotopological state, mean atomic van der Waals volume, and the total number of bonds as main influencing factors for the toxicity endpoint:

$$-\log LD_{50}\ (Oral\ rat,\ mol\ kg^{-1}) = 4.75 - 4.36Mv + 0.0196P\_VSAi\_1 - 0.236O - 058$$
$$+ 0.0224CATS2D\_03\_LL_{50} - 3.42ATSc2 - 0.0127nBondsS2$$
$$- 0.0677minssO + 0.937maxdssC$$

$$(3.1^*)$$

where definitions of the eight molecular descriptors appearing in eq. (3.1^*) are given in Table 3.1.

The reported values of LD_{50} are expressed as mg per kg bodyweight in literature, which indicates the dose of a substance killing 50% of tested animals after exposure by oral route. Sun et al. [208] converted all LD_{50} values (mg kg^{-1}) into a molar unit (mol kg^{-1}), followed by a negative logarithmic transformation ($-\log LD_{50}$ or pLD50). Since a higher value of $-\log LD_{50}$ value means higher toxicity, increasing the values of

https://doi.org/10.1515/9783111189673-003

Table 3.1: Definitions of the eight molecular descriptors appearing in eq. (**3.1***).

Descriptor	Definition	Type
maxdssC	Maximum atom-type E-state: =C<	Electrotopological state atom-type descriptor
P_VSA_i_1	P_VSA-like on ionization potential, bin 1	P_VSA-like descriptors (ionization potential)
CATS2D_03_LL	CATS2D lipophilic–lipophilic at lag 03	CATS 2D (basic descriptors)
Mv	Mean atomic van der Waals volume (scaled on carbon atom)	Constitutional indices (basic descriptors)
nBondsS2	Total number of single bonds	PaDEL bond count descriptor
ATSc2	Broto–Moreau autocorrelation descriptor of a topological structure, weighted by partial charges at lag 2 topological distance	Autocorrelation descriptor (centered Broto–Moreau autocorrelations)
O-058	= O group (sp^2 hybridization) in a molecule	Atom-centered fragments (basic descriptors)
minssO	Minimum atom-type E-state: -O-	Electrotopological state atom-type descriptor

descriptors *Mv, O-058, ATSc2, nBondsS2*, and *minssO* in eq. (**3.1***) can enhance the toxicity of PAHs.

3.2 Benzene Derivatives

Benzene derivatives have wide applications in herbicides, insecticides, and organic synthesis as solvents, propellants, cooling agents, additives, and other polymers [209]. Due to their toxicity, many of them can harm humans, the environment, animals, and plants [210]. Nitrobenzenes (NBs) are an important class of benzene derivatives, which have a significant potential for contamination [211]. They have wide applications including dye, medicine, and explosive synthesis [212]. Some QSAR/QSTPR models were introduced to express the toxicity of benzene derivatives and their subsets in literature [213–215]. Four important QSAR/QSTPR models to the ciliate *Tetrahymena pyriformis* for benzene derivatives are discussed in the following sections.

3.2.1 3D-QSAR/QSTPR Studies Using CoMFA, CoMSIA, and VolSurf Approaches

Salahinejad and Ghasemi [214] used 3D-QSAR/QSTR analysis to assess the toxicity of a large set of substituted benzenes toward ciliate *T. pyriformis*. They carried out the

3D-QSAR/QSTR studies using CoMFA, CoMSIA, and VolSurf techniques. They used toxicity data of 392 diverse structural substituted benzene containing nitrobenzenes, aminobenzenes, phenols, and benzonitrile to *T. pyriformis*. In addition to hydrophobic effects, the results of CoMFA and CoMSIA3-D contour maps confirmed that electrostatic and H-bonding interactions also play important roles in the toxicity of benzene. Salahinejad and Ghasemi [214] also introduced a fairly good predictive linear model based on VolSurf descriptors as follows:

$$-\log \text{IC}_{50}\,(\text{mM}) = -4.51 + 0.209 \log\ K_{\text{OW}} - 0.004 WN_1 + 0.012 W_3 + 1.369 CD_1$$
$$- 0.057 WN_5 + 2.747R + 0.687 SKIN \tag{3.2*}$$

where K_{OW} is the 1-octanol/water partition coefficient; WN_1 and WN_5 are two H bond acceptor volume descriptors; W_3 is a hydrophilic region descriptor; CD_1 is the ratio of the hydrophobic volume over the total molecular surface; R and $SKIN$ are rugosity and skin permeability, respectively. The value of log K_{OW} has a dominant role in the uptake of a molecule into or through biological membranes. The descriptor W_3 shows polarizability and dispersion forces, which were computed from molecular fields of -0.8 kcal mol^{-1}. The variable CD_1 accounts for the hydrophobic volume per surface unit calculated at an energy level of -0.2 kcal mol^{-1}. Since the mentioned descriptors could be useful in the description of dipole-induced dipole forces in a nonpolar molecule, they show the dominant and positive impact on toxicity. WN_1 and WN_5 represent the molecular envelope generating attractive H bond acceptor interactions, which could be computed from molecular fields at an energy level of -1 and -5 kcal mol^{-1}, respectively. Thus, they account for the polar and hydrogen bonding interactions in a polar molecule containing strong proton-transfer acids and base groups because the toxicity depends also on the propensity of forming hydrogen bonds and dipole–dipole forces [216]. The descriptor R measures the molecular wrinkled surface, which is computed as the ratio of volume/surface of a molecule [217]. Since cell membrane permeation is a prerequisite for the absorption and distribution of a compound in toxicity mechanism, $SKIN$ is defined as permeation through biological membranes [217].

3.2.2 Atom-Based Nonstochastic and Stochastic Linear Indices

TOpological MOlecular COMputer Design–Computer-Aided Rational Drug Design (TO-MOCOMD-CARDD) is a scheme for QSAR/QSTR studies [218]. Castillo-Garit et al. [219] developed two QSAR/QSTR for the prediction of aquatic toxicity of benzene derivatives using atom-based nonstochastic and stochastic linear indices using TOMOCOMD-CARDD software. They used data set toxicity data to the ciliate *T. pyriformis* for 392 benzene derivatives. They applied several steps to derive their QSAR/QSTR models with TOMOCOMD-CARDD software as follows:

1. The drawing mode is used to draw the molecular pseudograph for each molecule in the data set.
2. Appropriate weights were used to differentiate the molecular atoms for the calculation of the DRAGON descriptors [180, 220–222], which include atomic polarizability (P), atomic mass (M), van der Waals atomic volume (V), Mulliken atomic electronegativity (K), and the atomic electronegativity in Pauling scale (G) as shown in Table 3.2 [180, 220–222].

Table 3.2: Values of atomic weights used for linear indices calculation.

ID	Atomic mass (g mol⁻¹)	van der Waals volume (Å³)	Mulliken electronegativity	Polarizability (Å³)	Pauling electronegativity
H	1.01	6.709	2.592	0.667	2.20
B	10.81	17.875	2.275	3.030	2.04
C	12.01	22.449	2.746	1.760	2.55
N	14.01	15.599	3.194	1.100	3.04
O	16.00	11.494	3.654	0.802	3.44
F	19.00	9.203	4.000	0.557	3.98
P	30.97	26.522	2.515	3.630	2.19
S	32.07	24.429	2.957	2.900	2.58
Cl	35.45	23.228	3.475	2.180	3.16
Br	79.90	31.059	3.219	3.050	2.96
I	126.90	38.792	2.778	5.350	2.66

3. The total and local (atom and atom-type) atom-based linear indices of the molecular pseudograph's atom adjacency matrix are computed in the software calculation mode. The atomic properties and the descriptor family are calculated before calculating the molecular indices. A table is generated in which the rows correspond to the compounds, and the columns correspond to the atom-based (both total and local) linear maps or other molecular descriptor families implemented in this program.
4. A QSAR/QSTR equation is developed by the MLR method, for example, to find $-\log IC_{50}$ and the atom-based linear fingerprints having the following appearance:

$$-\log IC_{50}(mM) = c + a_0 f_0(x) + a_1 f_1(x) + a_2 f_2(x) + \cdots + a_k f_k(x) \tag{3.3}$$

where $f_k(x)$ is the kth total (atom and atom-type) linear indices, and the a_k and c are the coefficients obtained by MLR method.
5. The robustness and predictive power of eq. (3.3) are done by using internal and external (using an external prediction set) validation techniques.

Castillo-Garit et al. [219] computed the descriptors as follows:
(1) $f_k(x)$ and $f_k^H(x)$ are the kth atom-based nonstochastic total linear indices considering and not considering H atoms, respectively, in the molecule.

(2) $f_{kL}(x_E)$ and $f_{kL}^H(x_E)$ are the kth atom-based nonstochastic local (atom-type = hetero-atoms: S, N, O) linear indices considering and not considering H atoms, respectively, in the molecule.

(3) $f_{kL}^H(x_{E-H})$ are the kth atom-based nonstochastic local (atom-type = H atoms bonding to heteroatoms: S, N, O) linear indices considering H atoms in the molecular pseudograph ($G1$).

Castillo-Garit et al. [219] also computed the kth atom-based stochastic total [$^s f_k(x)$ and $^s f_k^H(x)$], as well as local [$^s f_k(x_E), ^s f_k^H(x_E)$ and $^s f_k^H(x_{E-H})$] linear indices. They introduced eqs. (3.4*) and (3.5*) as two statistically significant QSAR/QSTR models with nonstochastic ($R^2 = 0.791$) and stochastic ($R^2 = 0.799$) linear indices as follows:

(i) Atom-based nonstochastic linear indices:

$$-\log IC_{50}(mM) = -1.514 + 0.167^P f_{0L}^H(x_E) + 0.105^K f_0(x) - 0.360^P f_{1L}^H(x_E)$$
$$+ 0.0326^P f_{3L}^H(x_E) - 2.12 \times 10^{-3P} f_3^H(x) - 5.97 \times 10^{-7M} f_{9L}^H(x) \qquad (3.4^*)$$

(ii) Atom-based stochastic linear indices:

$$-\log IC_{50}(mM) = -1.471 + 0.0113^{Ms} f_{15}(x) + 1.242^{Ks} f_0^H(x) - 0.634^{Ks} f_5^H(x)$$
$$- 0.663^{Gs} f_0^H(x) - 0.0759^{Vs} f_{4L}(x_E) + 0.0795^{Vs} f_{2L}^H(x_E) \qquad (3.5^*)$$

Both eqs. (3.4*) and (3.5*) can be efficiently used to predict the aquatic toxicity of benzene derivatives but the performance of the stochastic model is better than that of the nonstochastic model.

3.2.3 Semiempirical Descriptors

Singh et al. [215] developed a QSAR/QSTR model for describing the toxicity of benzene derivatives against *T. pyriformis* using partition coefficient (K_{OW}) and semiempirical descriptors as follows:

$$-\log IC_{50}(mM) = 1.1878 + 0.5139 \log K_{OW} - 0.0003 TE + 0.0139 \varpi - 0.728 \eta \qquad (3.6^*)$$

where TE, ϖ, and η represent the total energy, the electrophilicity index, and the hardness, respectively, where details of these descriptors are given elsewhere [215]. Equation (3.6*) is a four-parameter model in which toxicity is positively influenced by $\log K_{OW}$ and ϖ and negatively influenced by TE and η.

Example 3.1: (a) Use eq. (**3.6***) for the calculation of $-\log IC_{50}(mM)$ of 1-phenyl-2-butanol. Assume the values of $\log K_{OW}$, TE, ϖ, and η are 2.02, −1,712.9392, 54.6471, and 4.8635, respectively. (b) If the experimental value of $-\log IC_{50}(mM)$ is −0.16 [215], calculate a deviation of (**3.6***).

Answer: (a) Equation (**3.6***) gives

$$-\log IC_{50}(mM) = 1.1878 + 0.5139 \log K_{OW} - 0.0003TE + 0.0139\varpi - 0.728\eta$$

$$= 1.1878 + 0.5139(2.02) - 0.0003(-1712.9392) + 0.0139(54.6471) - 0.728(4.8635)$$

$$= -0.04$$

(b) Dev = −0.04 − (−0.16) = 0.12

3.3 Phenol Derivatives

Some efforts have been done to predict the toxicity of phenol derivatives using different approaches such as MLR [223] and GA-PLS [224] with different sizes of the data set. Abbasitabar and Zare-Shahabadi [225] used a genetic algorithm and decision tree-based modeling approach to obtain a suitable QSAR/QSTR model for the prediction of toxicity of 206 phenols to *T. pyriformis* as follows:

$$-\log IC_{50}(mM) = 0.74 + 0.38 \log D + 0.19ATS\,3e - 0.33GATS1\,p - 0.13Mor23e$$

$$- 0.15R3u + + 0.37C - 026 \qquad (3.7^*)$$

where $\log D$ is the logarithm of the ionization-corrected octanol/water partition coefficient; $ATS\,3e$ is Broto–Moreau autocorrelation of lag 3 (log function)/weighted by Sanderson electronegativity; $GATS1\,p$ is Geary autocorrelation-lag 1/weighted by polarizability; $Mor23e$ is 3D-MoRSE-signal 23/weighted by Sanderson electronegativity; $R3u+$ is R maximal autocorrelation of lag 3/unweighted; $C-026$ is the number of atom-centered fragments R–CX–R. Table 3.3 defines the descriptors that appeared in the model. Among these descriptors, $ATS\,3e$ and $Mor23e$ are related to the electronegativity of the molecule. $GATS1\,p$ is dependent on the polarizability of the molecule. Two descriptors $R3u+$ and $C-026$ are related to the structural feature of the molecule. The positive sign of $C-026$ indicates that a branched chain comprising one or more heteroatoms enhances the toxicity. Toxicity is rather positively related to $\log D$, $ATS\,3e$, and $C-026$ because their coefficients in eq. (**3.7***) are positive. Due to the negative coefficients of $GATS1\,p$, $Mor23e$, and $R3u+$ in eq. (**3.7***), these descriptors have negative contributions to toxicity.

3.4 Benzoic Acid Derivatives

Benzoic acid derivatives are widely used as raw materials in medicine, food, cosmetics, antiseptic, insecticide, and dyestuff [226]. Many benzoic acid derivatives are toxic and hardly degraded by microorganisms in the natural environment [227]. Some works have been done to use QSAR/QSTR models for the prediction of toxicity of benzene derivatives [139, 228–232]. Several QSAR/QSTR models are reviewed here.

3.4.1 Predicting Toxicity Through Mouse via Oral LD_{50}

3.4.1.1 Modified Molecular Connectivity Index

Li et al. [230] calculated the quantum chemistry parameters of 57 benzoic acid derivatives with a modified molecular connectivity index (MCI) to derive a QSAR/QSTR model in mice via oral LD_{50} (acute toxicity). Their model is based on descriptors ${}^0 J^A$, ${}^1 J^A$, and cross-factor J^B as follows:

$$-\ln LD_{50}(Oral\ mouse, mg\ kg^{-1}) = -1.2399\,{}^0 J^A - 2.6911\,{}^1 J^A + 0.4445 J^B \qquad \textbf{(3.8*)}$$

where ${}^0 J^A$ is zero-order connectivity index; ${}^1 J^A$ is the first-order connectivity index; $J^B = {}^0 J^A \times {}^1 J^A$ is the cross-factor. As shown in eq. (**3.8***), ${}^0 J^A$, ${}^1 J^A$, and cross-factor J^B have a great influence on oral toxicity in mice. Since the coefficients of ${}^0 J^A$ and ${}^1 J^A$ have negative signs, increasing the values of ${}^0 J^A$ and ${}^1 J^A$ can increase the toxicity of a desired benzoic acid derivative. In contrast, decrement in the value cross-factor J^B may also enhance the toxicity of benzoic acid compounds.

3.4.1.2 Elemental Composition and Molecular Fragments

The success of QSAR/QSTR models depends on the accuracy of the input data, statistical tools, and selection of appropriate descriptors as well as their validation. Since eq. (**3.8***) requires special computer codes and complex descriptors, a simple model based on elemental composition and molecular fragments has been developed to estimate the toxicity of benzoic derivatives in mice via oral LD_{50} with 381 benzoic acid derivatives as follows:

$$-\ln LD_{50}(Oral\ mouse, mg\ kg^{-1}) = -8.485 + 0.2047 n_C - 0.1040 n_H - 2.554 ISM + 1.998 DSM$$
$$\textbf{(3.9*)}$$

where n_C and n_H are the number of carbon and hydrogen atoms in the molecular formula of benzoic acid compounds, respectively; *ISM* and *DSM* show increasing and decreasing structural moieties, respectively. The values of *ISM* and *DSM* can be specified according to the following situations:

(i) For the existence of –NH-C(=O)-C-C(=O)-NH-, –N-C(=O)-C-N-N-, -N-C(=O)-C(=O)-N, and –O-C(=O)-C-O-C(=O)-, *ISM* values are 1.4, 1.0, 0.8, and 0.4, respectively.

(ii) For the attachment of simultaneously only –I, -NH-C(=O)-CH$_3$ (or –N = N-), and –COOH to a benzene ring, the *ISM* value is 1.0.

(iii) For the attachment of ⬡N– from nitrogen to a benzene ring, *ISM* equals 0.3.

(iv) For the presence of a seven-membered ring ⬡ , the *ISM* value is 0.4.

(v) For the existence of –OH group ortho to –C(=O)-NH$_2$ in the benzene ring, the *DSM* value equals 0.7.

Example 3.2: (a) Calculate $-\ln LD_{50}$(*Oral mouse, mg kg*$^{-1}$) for the following molecular structure using eq. (**3.9***):

Chemical Formula: C$_9$H$_6$I$_3$NO$_3$

(b) Li et al. [230] calculated $-\ln LD_{50}$(*Oral mouse, mg kg*$^{-1}$)= –7.130 for this compound by eq. (**3.8***). If the experimental value of $-\ln LD_{50}$(*Oral mouse, mg kg*$^{-1}$) is –9.900 [233], compare the predicted results of eqs. (**3.8***) and (**3.9***) with the experimental data.

Answer: (a) The use of eq. (**3.9***) and condition (ii) with *ISM* = 1.0 gives

$$-\ln LD_{50}(\textit{Oral mouse, mg kg}^{-1}) = -8.485 + 0.2047n_C - 0.1040n_H - 2.554ISM + 1.998DSM$$

$$= -8.485 + 0.2047(9) - 0.1040(6) - 2.554(1.0) + 1.998(0) = -9.821$$

(b) Deviation of eq. (**3.8***) = –7.130 – (–9.900) = 2.770
Deviation of eq. (**3.9***) = –9.821 – (–9.900) = 0.079

Thus, the predicted result of eq. (**3.9***) is very close to the experimental data.

3.4.2 Estimating Toxicity Through Rats via Oral *LD$_{50}$*

3.4.2.1 The Effect of Quantum Chemistry Parameters

Sun et al. [139] introduced several QSAR/QSTR models to calculate the toxicity of benzoic acid derivatives in rats via oral *LD$_{50}$* using a computer program and the interactions of parameters. They used two-thirds of the 81 benzoic acid derivatives as a training data set to build models using the MLR method, and one-third of total compounds as a predicting test data set to test the model. The best predictive model for the prediction of toxicity of benzoic acid derivatives in rats via oral LD_{50} is given as follows:

$$-\ln LD_{50}(Oral\ rat,\ mg\ kg^{-1}) = -0.144\log K_{OW} + 0.0269 SAG - 1.27 \times 10^{-5}\Delta_f H(g)$$
$$+ 3.77 \times 10^{-4} PE1 \tag{3.10*}$$

where K_{OW} is octanol–water partition coefficient; SAG is surface area grid; $\Delta_f H(g)$ is gas-phase heat of formation, which is calculated by PM3 semiempirical quantum-mechanical method; $PE1$ = polarization energy × electronic energy. A few predicted values of eq. (3.10*) did not match the experiment values very well because (1) the experiment values were collected from different sources; (2) different species, gender, or quantity of rats may provide different experimental LD_{50} values; (3) the used descriptors in eq. (3.10*) mainly depend on the structure of compounds, without considering their particular environment.

3.4.2.2 Desk Calculation of Toxicity of Benzoic Acid Derivatives in Rats via Oral LD_{50}

A reliable correlation was introduced for desk calculation of toxicity of benzoic acid derivatives containing carbon, hydrogen, nitrogen, oxygen, and halogen atoms, or their sodium salts in rats via oral LD_{50} [232]. It is based on the number of nitrogen, oxygen, and halogen atoms as well as two correcting functions for increasing and decreasing the toxicity of benzoic derivatives. The predicted results of this model are more reliable than those obtained by Sun et al. [139] from eq. (8.10*) for 81 benzoic acid derivatives. The high reliability of the current model has also been checked for further 134 benzoic acid derivatives containing complex molecular structures. The correlation is given as follows:

$$-\ln LD_{50}(Oral\ rat,\ mg\ kg^{-1}) = -8.13 - 0.346 n_N + 0.149 n_O + 0.223 n_{Hal} + 1.53 C_{in} - 1.11 C_{de}$$
$$\tag{3.11*}$$

where n_N, n_O, and n_{Hal} are the number of nitrogen, oxygen (or sulfur), and halogen atoms in the molecular formula of a benzoic acid compound, respectively; C_{in} and C_{de} show the existence of specific molecular moieties for increasing and decreasing toxicity n_N based on n_O and n_{Hal}, respectively. The values of C_{in} and C_{de} can be found from the following molecular moieties:
(i) *Molecular moieties corresponding to C_{in}:*
   (a) For the presence of –OH, -OR, RCOO-, or –CH$_2$-CO- ortho to –CONH$_2$ or -COOH without the existence of –NH$_2$, -SO$_3$H, -Br as well as the presence of both –OH and –Cl simultaneously, C_{in}=1.0.
   (b) For the existence of –NH- or –N = C- between two benzene rings without the existence of –OH group, C_{in}=1.0.
   (c) For the presence of –N = N-N- and –SCN groups, C_{in} is 3.1 and 2.4, respectively.

(ii) *Molecular fragments corresponding to C_{de}:* For the existence of strong hydrogen bonding group para to –COOH in disubstituted benzene or the attachment of simultaneously –Cl, –NO$_2$, and –CO- groups to a benzene ring, C_{de}=1.0.

Equation (**3.11***) can be applied only for benzoic acid derivatives containing carbon, hydrogen, nitrogen, oxygen, and halogen atoms or their sodium salts. It cannot be used for some benzoic acid derivatives containing several specific molecular fragments, including -C = N-OH, -O-C(=O)-O-, –N = C(-S-)$_2$, and hydrochloride salts of benzoic acid compounds because the predicted results may show large deviations.

Example 3.3: (a) Use eq. (**3.11***) to calculate $-\ln LD_{50}(Oral\ rat,\ mg\ kg^{-1})$ for the following molecular structure:

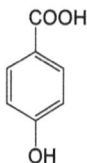

COOH

OH

Chemical Formula: $C_7H_6O_3$

(b) If the calculated and experimental values of $-\ln LD_{50}(Oral\ rat,\ mg\ kg^{-1})$, by the method of Sun et al. [139] using eq. (**3.10***), are −7.82 [233] and −9.21, respectively, compare the outputs of the two methods with the experimental data.

Answer: (a) The use of eq. (**3.11***) and condition (ii) with C_{de} = 1.0 gives

$$-\ln LD_{50}(Oral\ rat,\ mg\ kg^{-1}) = -8.13 - 0.346n_N + 0.149n_O + 0.223n_{Hal} + 1.53C_{in} - 1.11C_{de}$$
$$= -8.13 - 0.346(0) + 0.149(0) + 0.223(0) + 1.53(0) - 1.11(1) = -8.79$$

(b) Deviation of eq. (**3.10***) = −7.82 – (−9.21) = 1.39
Deviation of eq. (**3.11***) = −8.79 – (−9.21) = 0.42

Thus, the predicted result of eq. (**3.11***) is very close to the experimental data.

3.5 Assessment of Antitrypanosomal Activity of Sesquiterpene Lactones

Human African trypanosomiasis (HAT) or sleeping sickness is a protozoan-neglected tropical illness that arises in sub-Saharan Africa in which the insect vectors and tsetse flies of the *Glossina* species are endemic. It is the chief health anxiety in many countries in Africa [234]. It includes *Trypanosoma brucei gambiense* and *Trypanosoma bru-*

cei rhodesiense (Tbr) [235]. Due to high toxicity, high cost, trouble in management, and inaccessibility of drugs to resource-deprived rural groups, new harmless, active, and reasonable drugs are immediately desired [236]. Definite sesquiterpene lactones (STLs) are strong antitrypanosomal agents [237]. Several works were done to use QSAR works of STLs and their antitrypanosomal activity [238–241]. Kimani et al. [241] extended 130 STLs of various structural subclasses to apply 3 different QSAR approaches, which contain "classical" molecular descriptors, 3D pharmacophore features, and 2D molecular hologram QSAR (HQSAR). These approaches are complex and need expert users and complicated computer codes. Simple structural descriptors rather than complex descriptors have also been used to find the antitrypanosomal activity of STLs in terms of biological activity using the largest available experimental data sets as follows [242]:

$$-\log IC_{50}(Tbr, \ \mu M) = 4.0734 + 0.3536 n_{\text{Epoxy}} + 0.1764 n_{\text{Vinylic H except cyclo–CH}_2\text{CCOO}}$$

$$+ 1.0538 n_{\text{Cyclopentenone}} + 1.0538 n_{\text{Cyclopentenone}} + 1.0355 n_{\alpha-\text{CH}_2-\text{cyclo–COO}}$$

$$+ 0.1628 n_{\text{Acyclic–COO}} - 1.0748 pIC_{50}^{-} + 0.8084 pIC_{50}^{+}$$

$$(3.12^*)$$

where n_{Epoxy} is the number of epoxy groups; $n_{\text{Vinylic H except cyclo–CH}_2\text{CCOO}}$ is the number of vinylic hydrogen atoms, that is, the vinylic carbon atom is one of the two atoms that share the double bond, except two hydrogen atoms of vinylic carbon atom ($CH_2=$) in the alpha position of carbonyl group attached to a cyclic ring containing ester functional group; $n_{\text{Cyclopentenone}}$ is the number of cyclopentenone ring; $n_{\alpha-\text{CH}_2-\text{cyclo–COO}}$ is the number of alpha CH_2 = group attached to a cyclic ring containing ester functional group; $n_{\text{Acyclic–COO}}$ is the number of the acyclic ester group. Two parameters pIC_{50}^{-} and pIC_{50}^{+} show nonadditive influences of some specific groups in certain positions of the framework of different classes, which have been specified in Table 3.3.

Table 3.3: The values of pIC_{50}^- and pIC_{50}^+, where R represents the alkyl group.

Class	STL	Specific group	pIC_{50}^-	pIC_{50}^+	Example
1		X = -OH	0	1.0	
	or or		1.0	0	
			1.5	0	

2		X = CH₃CO	1.0	0	
3		X = O	1.5	0	
4		X = CHO	0	1.5	

(continued)

Table 3.3 (continued)

Class	STL	Specific group	pIC_{50}^-	pIC_{50}^+	Example
5		X and Y = R	0	1.5	
			0	1.0	
6		X = H	0	1.0	
			1.0	0	

Example 3.4: Use eq. (**3.12***) to calculate $-\log IC_{50}(Tbr, \mu M)$ for the following STL compound:

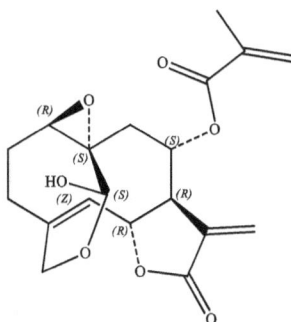

Answer: Since the values of n_{Epoxy}, $n_{Vinylic\ H\ except\ cylo-CH_2CCOO}$, $n_{a-CH_2-cyclo-COO}$, $n_{Cyclopentenone}$, and $n_{Acyclic-COO}$ are 1, 3, 0, 1, and 1, respectively, as well as pIC_{50}^{-} and pIC_{50}^{+} are zero, eq. (**3.12***) gives

$$-\log IC_{50}(Tbr, \mu M) = 4.0734 + 0.3536 n_{Epoxy} + 0.1764 n_{Vinylic\ H\ except\ cyclo-CH_2CCOO} + 1.0538 n_{Cyclopentenone}$$

$$+ 1.0355 n_{a-CH_2-cyclo-COO} + 0.1628 n_{Acyclic-COO} - 1.0748 pIC_{50}^{-} + 0.8084 pIC_{50}^{+}$$

$$-\log IC_{50}(Tbr, \mu M) = 4.0734 + 0.3536(1) + 0.1764(3) + 1.0538(0) + 1.0355(1) + 0.1628(1)$$

$$- 1.0748(0) + 0.8084(0) = 6.154$$

3.6 Assessment of Activities of Thrombin Inhibitors

Thrombi (blood clots) can form in blood vessels in thromboembolic disorders, which are among the main reasons for death in the world [243]. Vein thrombosis can improve pulmonary embolism (PE) because an embolus is a blood clot that travels through the bloodstream and blocks an artery. Deep vein thrombosis (DVT) can result from a blood clot in a deep vein, usually the leg, groin, or arm. PE happens when a DVT clot breaks free from a vein wall and travels to the lungs blocking some or all of the blood supply. Venous thromboembolism (VTE) is the sum of PE and DVT. Since VTE provides a dangerous and potentially deadly medical condition, it affects several million people around the world [244]. Thrombin is a trypsin-like serine protease. It has a major role in the processes of hemostasis and thrombosis. It is very attractive for antithrombotic therapy because it cleaves the protein fibrinogen at specific arginine residues to give polymerizable fibrin in the blood coagulation cascade [245]. Since an oral thrombin inhibitor (TI) is one of the important types of anticoagulants used in humans [246], small molecule derivatives of TIs can be introduced for antithrombotic therapy [247], for example, benzamide, chlorobenzene sulfonate, benzenesulfonate, acetamide, and benzimidamide [248]. Since these derivatives have alike frameworks or identical scaffolds with diverse substituents, they can offer different types of TIs with large differences in their thrombin inhibitory activities.

Several types of QSAR models such as CoMFA/CoMSIA have been used in pyrroloquinazolines as thrombin receptor antagonists [249]. Docking and molecular dynamics (MD) were used to investigate the structural features of several TIs [250]. The interactions can be specified between inhibitors and different subsites of the thrombin binding site. Since thrombin contains a wide variety of binding modes and geometries for the enzyme, most of the TIs bind to one or the other of the S1–S3 subsites of the active site and/or the fibrinogen recognition exosite. Nilsson et al. [251] introduced a QSAR model for designing a set of compounds to predict their binding constants. Bhunia et al. [252] used 3D-QSAR and MD simulations to contour structural determinants for the selectivity of illustrative various classes of thrombin-selective inhibitors. Mena-Ulecia et al. [250] considered the interactions of TIs with S1, S2, and S3 subsites of the thrombin binding site. They used docking and CoMSIA for 177 TIs as well as MD and free energy calculations. A simple approach was introduced for the prediction of inhibitory activities as $\log(10^3/K_i)$ values, where K_i is the inhibition constant [253]. Since the values of K_i are in nM, a simple correlation was given as follows [253]:

$$\log\left(10^3/K_i\right) = 0.1675 + 0.2752n_{\mathrm{N}} + 0.2179n_{\mathrm{Hal}} + 0.5272TI^+ - 1.2501TI^- \qquad (3.13^*)$$

where n_{N} and n_{Hal} are the number of nitrogen and halogen atoms, respectively; TI^+ and TI^- show nonadditive influences of some groups and molecular moieties, which have been specified in Table 3.4.

Table 3.4: The values of TI^+ and TI^-, where Ar represents aromatic compounds and R represents the alkyl group.

Class	Compound	Condition	TI^+	TI^-	Example
1		X = -OR, Y = -H, Q = -O-	2.1	0	
		X = Halogen or Ar-SO$_2$-, Y = -H, Q = -O-	2.7	0	
		X = Halogen, -CF$_3$, Y = -H, Q = =N-	1.0	0	
		X = R-SO$_2$-, Y = -H, Q = -O-	1.7	0	
		X = -N=, Y = =CH-, Q = -O- or = N-	1.5	0	
		X = -OR, Y = -H, Q = =N-	1.2	0	
		X = -CN, Y = -H, Q = -O-			
		X = -CH=, Y = =CH-, Q = -O-			
2		X = R-O- or R, Y = -CH$_2$-CH$_2$-CH$_2$- or --CH$_2$-C(=CH)-CH$_2$-, Q = -H	1.0	0	
		X = R, Y = -CH$_2$-CH$_2$-, Q = -H	0	1.1	
		X = R, Y = -CH$_2$-CH$_2$-CH$_2$-, Q = CH$_3$SO$_2$-	0	0.7	
		X = R, Y = -CH$_2$-CF$_2$-CH$_2$-, Q = H-			

(continued)

Table 3.4 (continued)

Class	Compound	Condition	TT^+	TT	Example
3		X = Halogen, Y = alkyl, Q =	2.1	0	
		X = Halogen, Y = RCOO-, Q =	0	1.4	
		X = Halogen, Y = RO-, Q =	0	1.2	
		X = RO- or CF$_3$, Y = Alkyl-, Q =	0	0	
		X = Halogen, Y = alkyl, Q =	0	1.0	
		X = Halogen, Y = -CH$_3$, Q =	0	0.8	
4		X = -OR or -CF$_3$, Y = -CH$_2$- or -S(O)$_2$-	0	1.2	

5

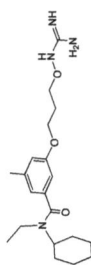

		1.7	0

X = Alkyl or cycloalkyl, Y = –CH₃, Q = alkyl or allyl

X =

, Y = –CH₃, Q = alkyl sulfide

X = , Y = Halogen or –CH₃, Q = alkyl except isopropyl and alkyl containing cyclobutyl and cyclopentyl — 1.2 0

X = Cycloalkyl, Y = –CH₃, Q = amide-containing alkyl chain — 1.0 0

X-N-Q = Cyclic ring, Y = halogen — 0 1.6

X = Alkyl, Y = halogen, Q = N

XNQ = Cyclic amine without substituent, Y = halogen — 0 1.1

X = Containing cyclic substituent, Y = halogen, Q = containing cyclic substituent — 0 0.9

X = Alkyl, Y = –CH₃, Q = — 0 0.7

6

X = –CH₃ — 2.1 0

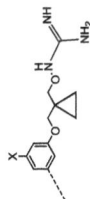

(continued)

Table 3.4 (continued)

Class	Compound	Condition	Π^+	Π	Example
7		X = −CN, Q = [structure], Ar = Aromatic compounds except for benzene, pyridine, methyl pyridine, chloropyridine oxide, and halopyridine	1.1	0	
		The existence of alkyl group attached to Ar in the ortho position of −CF₂− with the above substituents	0	0.7	
		X = Halogen or −CN, Q = pyridine derivatives	1.1	0	
		X = Halogen, Q = [structure]	0	1.3	
		Ar = aromatic compounds except for carbocyclic aromatic ring or bicycloaromatic ring such as quinolone or isoquinoline with its attachment through benzene ring as well as halopyridine			
		X = −CN	0	0.7	
		Q = [structure] or [structure]			

No.	Substituents		
8	X = Halogen, Y = -H, Q = -H, W = -CH₃	2.1	0
	X = RSO₂, Y = -H, Q = -H, W = -CH₃	1.0	0
	X = -NO₂, Y = -H, Q = -H, W = -CH₃		
	X = -H, Y = -R, Q = -H, W = -CH₃		
	X = -H, Y = -H, Q = -NH₂, W = -CH₃	0	0.7
	X = Halogen, Y = -H, Q = -H, W = -CH₂OH		
	X = -H, Y = -H, Q = -NO₂, W = -CH₃		
9	X = Halogen, Y = -H, Q = -H, W = -CH₃	0	0.6
10	Except for the presence of -CF₂-	0	1.0
11	Except for Ar = para-fluorobenzene or para-alkoxy benzene or naphthalene attached to -CH₂-	0	1.1
	Ar = The existence of -F and -CH₂- group in ortho- and para-positions, respectively, of -CH₂-	0	0.7

(continued)

Table 3.4 (continued)

Class	Compound	Condition	TT^+	TT	Example
12		X = Ar-SO$_2$- except ortho-substituent on Ar to -SO$_2$-, Y = ----N⟩ N-SO$_2$R	1.3	0	
		X = Benzene(derivatives)-SO$_2$- except ortho-substituent on benzene to -SO$_2$-, Y = ----N⟩ R	1.8	0	
		X = Polycyclic-CH$_2$-SO$_2$- or naphthalene (derivatives)-SO$_2$-, Y = ----N⟩ R	1.0	0	
			0	2.2	
		X = -Ar-SO$_2$-, Y = -NHR, -COO-, COO-,			
		X = -H, Y = ----N-SO$_2$R or ----N⟩ R	0	2.2	
		X = -Ar-SO$_2$-, Y = -NHR	0	1.6	
		X = Benzene (derivatives)-SO$_2$- with ortho-substituent on benzene to -SO$_2$-, Y = ----N⟩ CONH	0	1.5	

13

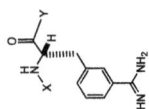

X = Benzene (derivatives)-SO$_2$- with ortho-substituent on benzene to –SO$_2$-,

Y =

| | 0 | 1.1 |

X = R-S(O$_2$)-,

Y =

| | 0 | 1.1 |

X = Benzene (derivatives)-SO$_2$- with ortho-substituent on benzene to –SO$_2$-,

Y =

| | 0 | 1.0 |

X = -Ar-SO$_2$-, Y =

| | 0 | 1.0 |

X =

Y =

| | 0 | 0.9 |

(continued)

Table 3.4 (continued)

Class	Compound	Condition	Π^+	Π^-	Example
		X = -Ar-SO$_{2^-}$ except benzene (derivatives)-SO$_{2^-}$ with ortho-substituent on benzene to -SO$_{2^-}$, 	0		0.7
		X = -Ar-SO$_{2^-}$ except benzene (derivatives)-SO$_{2^-}$ with ortho-substituent on benzene to -SO$_{2^-}$, 	0		0.5

Example 3.5: (a) Use eq. (**3.13***) to calculate $\log(10^3/K_i)$ for the following TI compound:

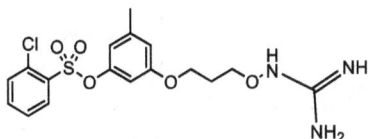

Chemical Formula: $C_{17}H_{20}ClN_3O_5S$

(b) If the measured $\log(10^3/K_i)$ is 2.68, compare the reliability of the predicted result of eq. (**3.13***) with that computed by Mena-Ulecia et al. [250], that is, 2.12.

Answer: (a) Equation (**3.13***) and Table 3.4 (Class 2, item 2) give

$$\log(10^3/K_i) = 0.1675 + 0.2752n_N + 0.2179n_{Hal} + 0.5272TI^+ - 1.2501TI^-$$
$$= 0.1675 + 0.2752(3) + 0.2179(1) + 0.5272(2.7) - 1.2501(0) = 2.63$$

(b) Equation (**3.13***): Dev = −0.05
    Mena-Ulecia et al.: Dev = 0.56

Thus, eq. (**3.13***) gives a more reliable result.

3.7 Assessing the Psychotomimetic Activity of the Substituted Phenethylamines

Psychedelics or hallucinogens can yield unique and dramatic alterations in consciousness. Since psychedelic drugs such as mescaline, psilocybin, and lysergic acid diethylamide have powerful effects on the human psyche, they provide such profound effects on perception [254]. Hallucinogens of phenethylamine give characteristic changes in human awareness, giving rise to a variety of abnormal phenomena. The substituted phenethylamines are the most extensively explored class of psychedelics largely because of the relatively superficial synthesis of phenethylamines. Mescaline or 2-(3,4,5-trimethoxyphenyl) ethanamine is a typical derivative of this category which is an orally active hallucinogen in humans. Due to the very low potency of mescaline, the structure of mescaline can be modified for the introduction of a very large class generically referred to as "substituted amphetamine" hallucinogens, for example, moving the methoxy ring substituents to diverse positions may give isomers with different potency. The most potent hallucinogenic amphetamines have the 2,4,5-ring-substitution pattern. Active compounds among a large number of substituted phenethylamines generally have a great affinity and are agonists or partial agonists at the 5-HT2A receptor. A chiral center is generated by the introduction of the α-methyl into the phenethylamine side chain, which can yield the substituted amphetamine-type psychedelics with two optical isomers, or enantiomers.

The psychotomimetic activity of psychedelic drugs is given in terms of standard mescaline [255]. Some studies have been done to forecast the activities of the substituted phenethylamines because they can display psychotomimetic activity. Because of the limitations and difficulty of psychotomimetic activity on the human being, it is appreciated to have reliable predictive methods for the valuation of the psychotomimetic activity of the desired compounds. The available QSAR models require physicochemical as well as electronic and charge descriptors [256–258], for example, a quantum-mechanical DFT-B3LYP method with the base set 6–31G (d) for calculation of electronic and charge descriptors [256]. Aouidate et al. [256] introduced QSAR models based on experimental data of 46 substituted phenethylamines for the prediction of their psychotomimetic activity. Two QSAR MLR and nonlinear multiple regression analysis (NMLR) models of Aouidate et al. [256] are given as follows:

MLR: $\log A = 10.99 - 2.61 \times 10^{-5} E_T + 1.29 E_{\text{HOMO}} - 1.88 E_{\text{LUMO}} - 0.29\mu - 6.60 \times 10^{-2}\gamma$ (**3.14***)

NMLR: $\log A = 88.55 - 8.11 \times 10^{-5} E_T + 22.17 E_{\text{HOMO}} - 3.69 E_{\text{LUMO}} - 0.35\mu - 1.27\gamma$

$$- 5.37 \times 10^{-10} E_T^2 + 1.97 E_{\text{HOMO}}^2 + 6.23 E_{\text{LUMO}}^2 + 4.87 \times 10^{-3}\mu^2 + 0.05\gamma^2$$

(**3.15***)

where $\log A$ is the psychotomimetic activity; E_{HOMO}, E_T, E_{LUMO}, μ, and γ are the highest occupied molecular orbital energy, the total energy, the lowest unoccupied molecular orbital energy, the dipole moment, and the surface tension, respectively.

Since the used descriptors in available QSAR models [256–258] are complex, it is valuable to correlate the psychotomimetic activity of the substituted phenethylamines and their structural parameters. A simple method has been introduced for calculating the psychotomimetic activity of the substituted phenethylamines without using complex computer codes and unusual descriptors as follows [259]:

$$\log A = 0.270 + 0.2731 C_{Iso} + 0.6229 C_{Hal}$$ (**3.16***)

where $\log A$ is the psychotomimetic activity; two variables C_{Iso} and C_{Hal} are the contributions of specific isomers under certain conditions and halogen atoms, respectively. The value C_{Hal} is equal to 1.0 for the presence of halogen atoms in the substituted phenethylamines. Table 3.5 gives different values of C_{Iso}.

Table 3.5: The values of C_{Iso}.

Class	Structural isomers	Condition			C_{Iso}
1		Y = –H	X =	Halogen	4.7
				Nitro	6.0
				Alkyl	5.5
				Thioalkyl	5.0
				Alkoxy	2.5
		(a) X and Y form cyclic alkane or (b) both X and Y are alkyl groups			3.5
		X = alkyl and Y = thioalkyl			3.0
		(a) X and Y form cyclic ether or (b) both X and Y are alkoxy groups			2.0
2		X = Alkyl			1.0
3		X =		Halogen	2.5
				Alkoxy	1.0
4		X, Y, or Z =		Alkoxy	2.5
				Alkyl	3.5
5		Y = –H and Z = -CH$_2$-CH$_2$- NH$_2$ or –CH(OCH$_3$)-CH $_2$-NH$_2$ or	X =	(a) Alkyl or (b) X and Y form cyclic alkane or (c) Both X and Y are alkyl groups	3.5
				Thioalkyl	2.5

Table 3.5 (continued)

Class	Structural isomers		Condition		C_{Iso}
6		R = Alkyl	X = Alkoxy	Y = Alkoxy	−1.0
				Y = -H	−3.5
				Y = Thioalkyl	−2.0
			X = Thioalkyl	Y = Methoxy	1.0
				Y = Alkoxy except methoxy	−2.0
		R = Methyl	X = Alkoxy	Y = Thioalkyl	1.0
				Y = Methoxy	2.0
			X = Thioalkyl	Y = Methoxy	3.0

Example 3.6: Consider 2-(2,3,4-trimethoxyphenyl)ethanamine with the following molecular structure:

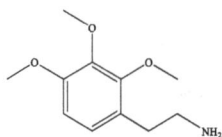

(a) Using eqs. (**3.14***), (**3.15***), and (**3.16***), calculate log A. Assume that E_T, E_{HOMO}, E_{LUMO}, DM, and y are −19,312.8, −5.719, 0.288, 1.449, and 35.5, respectively. (b) If the reported value of log A is −0.030 [255], predict a closer result as compared to the experimental data.

Answer: Since this compound has no contribution of C_{Iso} and C_{Hal}, eq. (**3.16***) gives log $A = 0.270$. Meanwhile, eq. (**3.14***) provides the value of log A as

$$\log A = 10.99 - 2.61 \times 10^{-5} E_T + 1.29 E_{HOMO} - 1.88 E_{LUMO} - 0.29 DM - 6.60 \times 10^{-2} y$$

$$= 10.99 - 2.61 \times 10^{-5}(-19,312.8) + 1.29(-5.719) - 1.88(0.288) - 0.29(1.449) - 6.60 \times 10^{-2}(35.4)$$

$$= 0.84$$

But eq. (**3.15***) gives

$$\log A = 88.55 - 8.11 \times 10^{-5} E_T + 22.17 E_{HOMO} - 3.69 E_{LUMO} - 0.35 DM - 1.27 y$$

$$-5.37 \times 10^{-10} E_T^2 + 1.97 E_{HOMO}^2 + 6.23 E_{LUMO}^2 + 4.87 \times 10^{-3} DM^2 + 0.05 y^2$$

$$= 88.55 - 8.11 \times 10^{-5}(-19,312.8) + 22.17(-5.719) - 3.69(0.288) - 0.35(1.449) - 1.27(35.4)$$

$$-5.37 \times 10^{-10}(-19,312.8)^2 + 1.97(-5.719)^2 + 6.23(0.288)^2 + 4.87 \times 10^{-3}(1.449)^2 + 0.05(35.4)^2$$

$$= 0.72$$

Since the reported value of log A is −0.030 [255], eq. (**3.16***) gives a closer prediction.

Example 3.7: Consider 2-(3,4-dimethoxyphenyl)ethanamine with the following molecular structure:

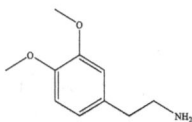

(a) Use eqs. (**3.14***), (**3.15***), and (**3.16***) to calculate log A. Assume that E_T, E_{HOMO}, E_{LUMO}, DM, and y are $-16,196.9$, -5.443, 0.406, 3.724, and 36.0, respectively. (b) If the reported value of log A is -0.67 [255], predict a closer result as compared to the experimental data.

Answer: According to the sixth class given in Table 3.5, the value of C_{Iso} is -3.5 without the contribution of C_{Hal}. Thus, eq. (**3.16***) gives

$$\log A = 0.270 + 0.2731 C_{Iso} + 0.6229 C_{Hal} = 0.270 + 0.2731(-3.5) + 0.6229(0) = -0.686$$

Meanwhile, eqs. (**3.14***) and (**3.15***) provide the values of log A as follows, respectively:

$$\log A = 10.99 - 2.61 \times 10^{-5} E_T + 1.29 E_{HOMO} - 1.88 E_{LUMO} - 0.29 DM - 6.60 \times 10^{-2} y$$

$$= 10.99 - 2.61 \times 10^{-5}(-16,196.9) + 1.29(-5.443) - 1.88(0.406) - 0.29(3.724) - 6.60 \times 10^{-2}(36)$$

$$= 0.20$$

$$\log A = 88.55 - 8.11 \times 10^{-5} E_T + 22.17 E_{HOMO} - 3.69 E_{LUMO} - 0.35 DM - 1.27 y$$

$$-5.37 \times 10^{-10} E_T^2 + 1.97 E_{HOMO}^2 + 6.23 E_{LUMO}^2 + 4.87 \times 10^{-3} DM^2 + 0.05 y^2$$

$$= 88.55 - 8.11 \times 10^{-5}(-16,196.9) + 22.17(-5.443) - 3.69(0.406) - 0.35(3.724) - 1.27(36)$$

$$-5.37 \times 10^{-10}(-16,196.9)^2 + 1.97(-5.443)^2 + 6.23(0.406)^2 + 4.87 \times 10^{-3}(3.724)^2 + 0.05(36)^2$$

$$= 0.19$$

Equation (**3.16***) gives a closer prediction because the experimental value of log A is -0.67 [255].

3.8 Summary

Equation (**3.1***) demonstrated a suitable model for the rat's acute oral toxicity (LD_{50}) of PAHs. Equations (**3.2***), (**3.4***), (**3.5***), and (**3.6***) can predict the toxicity of benzene derivatives to the ciliate *T. pyriformis*. A suitable QSAR/QSTR model for the prediction of toxicity of phenol derivatives to *T. pyriformis* was discussed in Section 3.3. Four models were demonstrated in Section 3.4 for the prediction of toxicity of benzoic acid derivatives. Equations (**3.8***) and (**3.9***) were based on the estimation of toxicity of benzoic derivatives in mice via oral LD_{50} for benzoic acid derivatives. Meanwhile, eqs. (**3.10***) and (**3.11***) can predict the toxicity of benzoic acid derivatives in rats via oral LD_{50}. In contrast to eqs. (**3.8***) and (**3.10***), eqs. (**3.9***) and (**3.11***) were based on simple structural parameters, which can be easily applied to the valuation of toxicity of benzoic acid derivatives. Finally, several approaches were introduced for assessments of the antitrypa-nosomal activity of STLs, and activities of thrombin inhibitors, and the psychotomimetic activity of the substituted phenethylamines. Equation (**3.12***) introduced the use of simple structural descriptors rather than complex descriptors to find the antitrypanosomal activity of STLs in terms of biological activity. Equation (**3.13***) presented a simple approach for the prediction of inhibitory activities as $\log(10^3/K_i)$ values to consider the interactions of TIs with S1, S2, and S3 subsites of the thrombin binding site. Equations (**3.14***), (**3.15***), and (**3.16***) were introduced for calculating the psychotomimetic activity of the substituted phenethylamines. In comparison to eqs. (**3.14***) and (**3.15***), eq. (**3.16***) did not require complex and unusual descriptors.

Problems

1. (a) Use eq. (**3.6***) for the calculation of $-\log IC_{50}(mM)$ of 5-phenyl-1-pentanol. Assume the values of $\log K_{OW}$, TE, ϖ, and η are 2.77, $-1,862.70222$, 51.7554, and 4.8865, respectively. (b) If the experimental value of $-\log IC_{50}$ (mM) is 0.42 [215], calculate a deviation of (**3.6***).

2. (a) Calculate $-\ln LD_{50}(Oral\ mouse,\ mg\ kg^{-1})$ for the following molecular structure using eq. (**3.9***):

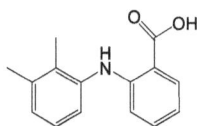

Chemical Formula: $C_{15}H_{15}NO_2$

(b) Li et al. [230] calculated $-\ln LD_{50}(Oral\ mouse,\ mg\ kg^{-1})= -3.290$ for this compound by eq. (**3.8***). If the experimental value of $-\ln LD_{50}(Oral\ mouse,\ mg\ kg^{-1})$ is -6.260 [233], compare the predicted results of eqs. (**3.8***) and (**3.9***) with the experimental data.

3. (a) Use eq. (**3.11***) to calculate $-\ln LD_{50}(Oral\ rat,\ mg\ kg^{-1})$ for the following molecular structure:

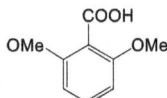

Chemical Formula: $C_9H_{10}O_4$

(b) If the calculated and experimental values of $-\ln LD_{50}(Oral\ rat,\ mg\ kg^{-1})$, by the method of Sun et al. [139] using eq. (**3.10***), are -7.49 [233] and -6.21, respectively, compare the outputs of the two methods with the experimental data.

4. Use eq. (**3.12***) to calculate $-\log IC_{50}(Tbr,\ \mu M)$ for the following STL compound:

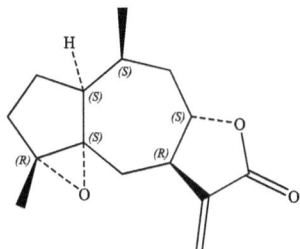

5. (a) Use eq. (**3.13***) to calculate $\log(10^3/K_i)$ for the following TI compound:

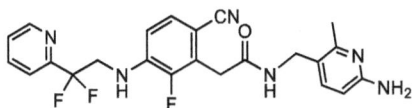

Chemical Formula: $C_{23}H_{21}F_3N_6O$

(b) If the measured $\log(10^3/K_i)$ is 3.11, compare the reliability of the predicted result of eq. (**3.13***) with that computed by Mena-Ulecia et al. [250], that is, 2.87.

6. Consider 1-(4-bromo-2,5dimethoxyphenyl) propan-2-amine with the following molecular structure:

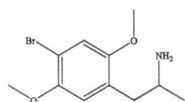

(a) Use eqs. (**3.14***), (**3.15***), and (**3.16***) and calculate $\log A$. Assume E_T, E_{HOMO}, E_{LUMO}, DM, and γ are −82,328.726, −5.682, −0.160, 3.242, and 37.9, respectively.
(b) If the reported value of $\log A$ is 2.176 [260], predict a closer result as compared to the experimental data.

Chapter 4
Toxicity of Large Data Sets

Assessments of toxicities of aromatic and organic compounds are important because they provide general models, which can be applied to many compounds including those compounds given in previous chapters. Since experimental data size for toxicity depends on the kind of protozoa, this chapter demonstrates general methods for aromatic and organic compounds.

4.1 Aromatic Compounds

Aromatic compounds are produced in large quantities and released into the environment as a result of their wide use in agriculture and industry, and are widely distributed in air, natural water, wastewater, soil, sediment, and living organics [261]. They are also a kind of biotoxic environmental pollutant and even have the effects of carcinogenesis and gene mutation on organisms [262]. They may be used for preparing high-octane fuel and are also used for a variety of other chemicals like gasoline, chemical intermediate, paints, lacquers, and dyes. Due to their low reactivity and smell enhancement, they are also used as solvents. They can cause aplastic anemia, lymphatic cancers, chronic lymphocytic leukemia, hematopoietic, excessive bleeding, and damage to the immune system. Aromatic compounds such as phenol derivatives can be solubilized with water to a greater extent, which causes serious problems for aquatic animals and the surrounding environment. Thus, it is necessary to estimate the toxicity of aromatic compounds such as benzene, phenol, nitrobenzene, and aromatic aldehydes before coming to the market for use, which can risk the life of human beings as well as other environmental habitats. Several new quantitative structure–activity/toxicity relationships (QSAR/QSTR) models are introduced for aromatic compounds here.

4.1.1 Regression-Based QSTR and Read-Across Algorithm

Kumar et al. [263] used a partial least-squares method (PLS) regression approach to develop QSAR/QSTR models for the prediction of the toxicity of *Tetrahymena pyriformis* using simple and easily interpretable 2D descriptors employing a data set containing 892 chemicals. They used the "Intelligent Consensus Predictor" tool and Read-Across software to show better results for test set compounds as compared with individual PLS models. They also validated the models using a set of 383 external set compounds, which have not been used for the development of their models. They developed five PLS models with a defined endpoint ($-\log IC_{50}(mM)$) against *T. pyriformis* employing a reduced pool of descriptors as follows:

https://doi.org/10.1515/9783111189673-004

Model M1

$$-\log IC_{50}(mM) = -2.034 - 0.1157H - 051 - 0.8426ESOL + 0.01395TPSA(NO) + 0.2867mindsN$$
$$- 0.5270n_{ROH} + 0.09101SssCH2 - 0.09209F03[C\text{-}N] - 0.06893T(Br \ldots Br)$$

$$(4.1^*)$$

Model M2

$$-\log IC_{50}(mM) = -2.085 + 1.242n_{RCHO} - 0.8431ESOL + 0.01454TPSA(NO) + 0.3203mindsN$$
$$- 0.5305n_{ROH} + 0.1021SssCH2 - 0.09619F03[C\text{-}N] - 0.06706T(Br \ldots Br)$$

$$(4.2^*)$$

Model M3

$$-\log IC_{50}(mM) = -2.042 + 1.259n_{RCHO} - 0.8354ESOL + 0.01313TPSA(NO) + 0.3028mindsN$$
$$- 0.5000n_{ROH} + 0.1012SssCH2 - 0.1102F04[C\text{-}N] - 0.0641T(Br \ldots Br)$$

$$(4.3^*)$$

Model M4

$$-\log IC_{50}(mM) = -1.941 - 0.1117H - 051 - 0.8147ESOL + 0.01289TPSA(NO) + 0.3015mindsN$$
$$- 0.5459n_{ROH} + 0.1027SssCH2 - 0.1185F04[C\text{-}N] - 0.06387T(Br \ldots Br)$$

$$(4.4^*)$$

Model M5:

$$-\log IC_{50}(mM) = -1.960 - 0.1404H - 051 + 0.4542\log K_{OW,WC} + 0.01012SAacc + 0.5443mindddC$$
$$- 0.6220O - 056 + 1.262n_{RCHO} - 0.3166BLTD48 - 0.07639F04[C\text{-}N]$$

$$(4.5^*)$$

The values of R^2 for the five models are in the range of 0.737–0.740. The description and contribution of the descriptors in eqs. (4.1*)–(4.5*) are given in Table 4.1.

As shown in Table 4.1, descriptors contribute negatively or positively toward *T. pyriformis*, which can be described in the following sections.

4.1.1.1 Descriptors with Negative Contributions

Descriptor ESOL is the most important descriptor, which indicates the estimated solubility for aqueous solubility because increasingly the solubility of a compound in an aqueous can easily be solubilized with water. Thus, excretion through the body can decrease toxicity. Since this descriptor has a negative regression coefficient, the presence of this fragment may decrease the toxicity profile of aromatic compounds toward *T. pyriformis*.

Table 4.1: Description and contribution of the descriptors in eqs. (**4.1***)–(**4.5***).

Descriptor	Description	Types of descriptor	Contribution	Fragment	Mechanistic interpretation
mindsN	Minimum dsN	Atom-type E- state indices	Positive	–N =	Increasing the value of the *mindsN* fragment will make the compound more toxic due to its positive contribution
nRCHO	Number of aldehydes (aliphatic)	Functional group counts	Positive		The presence of the *nRCHO* fragment increases the toxicity of compounds by forming adducts
O-056	Alcohol	Atom-centered fragments	Negative	OH	The presence of alcoholic fragments in aromatic structure leads to a decrease in toxicity by oxidizing to benzoic acid, conjugated with glycine in the liver, and excreted as hippuric acid
SAacc	Surface area of acceptor atoms from P_VSA-like descriptors	Molecular properties	Positive	–	*SAacc* fragment showing a positive contribution toward the toxicity of aromatic organic compounds
minddC	Minimum ddC	Atom-type state indices	Positive	=C =	*minddC* fragment contributes to positive toxicity of aromatic organic compounds
F04[C–N]	Frequency of C–N at topological distance 4	2D atom pairs	Negative		The presence and increase of the *F04[C-N]* fragment make compounds less toxic by reducing the hydrophobicity of the compound

Table 4.1 (continued)

Descriptor	Description	Types of descriptor	Contribution	Fragment	Mechanistic interpretation
H-051	H attached to alpha-C	Atom-centered fragments	Negative	H-	The presence and increase of the *H-051* fragment make compounds less toxic due to an increase in steric hindrance at alpha carbon
ESOL	Estimated solubility (log S) for aqueous solubility using log K_{ow} (LOGPcons)	Molecular properties	Negative	–	The presence of *ESOL* fragments make compounds less toxic due to an increase in aqueous solubility
TPSA(NO)	Topological polar surface area using N,O polar contributions	Molecular properties	Positive	–	The presence of *TPSA*(NO) fragments make compounds more toxic by enhancing hydrogen bonding
BLTD48	Verhaar *Daphnia* baseline toxicity from Moriguchi octanol–water partition coefficient (log $K_{ow,M}$)	Molecular properties	Negative	–	The toxicity of the compound decreases with the increase of *BLTD48* fragments in aromatic organic compounds
n_{ROH}	Number of hydroxyl groups	Functional group counts	Negative	HO-R	The presence of n_{ROH} fragments makes compounds less toxic due to the weakening of the interaction between aromatic organic compounds and *T. pyriformis*
$SssCH_2$	Sum of ssCH$_2$ E-states	Atom-type E-state indices	Positive	-CH$_2$-	The presence of *SssCH$_2$* fragment makes compounds more toxic by reducing the hydrophilicity of the compound

Table 4.1 (continued)

Descriptor	Description	Types of descriptor	Contribution	Fragment	Mechanistic interpretation
F03[C–N]	Frequency of C–N at topological distance 3	2D atom pairs	Negative	(structure with NH_2)	The presence of C and N atoms at topological distance 3 makes the compounds less toxic by reducing the hydrophobicity of the compound
T(Br. . .Br)	Sum of topological distances between Br . . .Br	2D atom pairs	Negative	(structure with Br substituents)	The presence of the *T(Br . . .Br)* fragment makes compounds less toxic due to a decrease in π–π interaction
log K$_{OW,WC}$	Logarithm of Wildman–Crippen octanol–water partition coefficient (log K_{OW})	Molecular properties	Positive	–	The higher value of log $K_{OW,WC}$ (more lipophilicity contributing fragments) makes compounds more toxic due to an increasing tendency of binding to unwanted cellular targets

Descriptor n_{ROH} is the next important descriptor, which counts the number of hydroxyl groups present in the compounds. The hydroxyl group cannot donate the lone pair of electrons to the aromatic rings. Since it weakens the interaction between the aromatic organic compound and *T. pyriformis*, this descriptor with a negative coefficient decreases the toxicity profile of aromatic compounds toward *T. pyriformis*.

Descriptor *O-056* is the atom-centered fragment descriptor, which shows the presence of alcoholic groups in the molecules. The existence of -OH groups in aromatic structure can decrease its toxicity because this compound may oxidize to benzoic acid. Benzoic acid can conjugate with glycine in the liver and excreted as hippuric acid [264]. Thus, the presence of a negative regression coefficient of *O-056* decreases the value of –log IC_{50} of an aromatic compound toward *T. pyriformis*.

Descriptor *BLTD48* is a molecular properties descriptor that represents the Verhaar *Daphnia* baseline toxicity from the Moriguchi octanol–water partition coefficient (log $K_{OW,M}$). Since it has a negative regression coefficient, *BLTD48* is inversely proportional to the toxicity of organic aromatic compounds against *T. pyriformis*.

Descriptors *F04[C–N]* and *F03[C–N]* denote the frequency of the C–N fragment at topological distances 4 and 3, respectively. Descriptor *T(Br.Br)* is the least important descriptor, which is defined as the sum of topological distances between two bromine atoms. The presence of polar group C–N in the case of *F04[C–N]* and *F03[C–N]* as well as two bromine atoms in the case of *T(Br.Br)* makes the aromatic compounds hydrophilic. Since these descriptors have negative regression coefficients, their presence makes the organic aromatic compounds more hydrophilic and less toxic to *T. pyriformis*.

Descriptor *H-051* is an atom-centered fragment descriptor, which describes the H atom attached to alpha-C. It has the negative regression coefficient that this descriptor is inversely proportional to the toxicity of aromatic compounds against *T. pyriformis*.

4.1.1.2 Descriptors with Positive Contributions

Descriptor $\log K_{OW,WC}$ is a molecular property descriptor, which is the Wildman–Crippen octanol–water partition coefficient to express the lipophilicity of a molecule [265]. The high value of the partition coefficient for a compound tends to its accumulation in the fatty tissue of the organisms (bioaccumulation). Thus, the highly lipophilic compounds will be more toxic against *T. pyriformis*, which can increase the tendency of binding to unwanted cellular targets and lead to toxicity [266].

Descriptor n_{RCHO} represents the number of aldehydes (aliphatic) present in the chemicals. Since it has a positive regression coefficient, it increases the values of –log IC_{50} of aromatic compounds to *T. pyriformis*. Aliphatic aldehyde groups in a molecule can increase its toxicity due to the formation of adducts by nucleophilic residues on *T. pyriformis*. Since enzyme active sites of *T. pyriformis* may contain cysteine residues, they can act as a nucleophile for soft unsaturated aldehyde group presence in aromatic aldehyde compounds. Nitrogen atoms of deoxyguanosine or amino group on lysine residues of *T. pyriformis* are frequently found as an adduct with hard alkanals of aromatic aldehyde compounds. These soft–soft and hard–hard adduct reactions can involve toxicity by impairing the function of *T. pyriformis* such as proteins, DNA, and RNA [267].

Descriptor SAacc is defined as the surface area of acceptor atoms from *P_VSA*-like descriptors, which show the amount of van der Waals surface area (*VSA*) having a property *P* in a certain range. Thus, *P_VSA* descriptors are calculated from approximate *VSA* using connection table approximation and some atomic property *P*. The interaction between aromatic organic chemicals and *T. pyriformis* increases with the increment of the *VSA*. Since it remains for a long time in *T. pyriformis* to increase the toxicity of a molecule [267], the toxicity of a molecule against *T. pyriformis* will be increased with the increment of this descriptor.

Descriptor *minddC* is explained as the minimum atom-type E-state: = C = , which is an electrotopological (*E*)-state atom indices descriptor. The minimum value of the minddC descriptor depends on the presence of methylene (=C=) groups. Since minddC descriptor has a positive regression coefficient, the higher numerical value of this descriptor is more toxic against *T. pyriformis*.

Descriptor *mindsN* is another electrotopological (*E*)-state atom indices descriptor, which indicates a minimum of E-state values for the –N= fragment. Thus, the presence of –N= fragment in an aromatic compound can increase its toxicity against *T. pyriformis*.

Moreover, decreasing the numerical value of this descriptor in aromatic compounds may decrease their toxicity.

Descriptor $SssCH_2$ is the atom-type E-state indices descriptor, which is defined as the sum of E-state values of the fragment of the methylene group. The atom-type E-state indices can be calculated as an average of the E-state values of all atoms of a given atom type within the molecule. They may also be calculated by adding the E-state values of all atoms of a given atom type together with the molecule. Due to a positive regression coefficient of this descriptor, the presence of this fragment makes organic chemicals more toxic.

Descriptor *TPSA(NO)* indicates the polar contribution of the topological polar surface area using nitrogen and oxygen atoms. The existence of both oxygen and nitrogen atoms in the same molecules can increase the overall electronegativity of the compound which is followed by oxidative stress, and death of the reference species. A positive regression coefficient of this descriptor shows that the presence of more electronegative atoms makes aromatic organic compounds more favorable to enhance toxicity by entering into the system of *T. pyriformis*.

4.1.2 Acute Toxicity of Aromatic Chemicals in Tadpoles of the Japanese Brown Frog (*Rana japonica*) Using Correlation Weights

The vertebrate class Amphibia including Anura (frogs, toads, and relatives), Caudata (salamanders, newts, and relatives), and *Gymnophiona* (caecilians and relatives) constitutes an important taxonomical group for which all modern amphibians belong to the subclass *Lissamphibia* [268]. Amphibian populations are undergoing a global decline in numbers, which have been attributed to their unique physiology, ecology, and exposure to multiple stressors including chemicals, temperature, biological hazards such as fungi of the *Batrachochytrium* genus, viruses such as *Ranavirus*, and habitat reduction [269].

For substances such as plant protection products and environmental contaminants, ecotoxicological studies are still limited in amphibian species investigating chemical toxicity because historically more focus has been given to aquatic vertebrates such as fish test species rainbow trout and zebrafish [270]. There is a lack of experimental data for the different life stages of amphibians and regulatory legislation requiring environmental risk assessment (ERA) of chemicals in amphibians. European directives for industrial chemicals and plant protection products require data on aquatic organisms such as insects, fish, daphnia, and algae, but not amphibians. The European Food Safety Authority (EFSA) published a scientific opinion on the state of the science on pesticide ERA for amphibians and reptiles. It provides a scientific basis addressing their sensitivity to pesticides as well as data gaps and formulates recommendations for further support of the inclusion of these species are given in ERA [270].

Since amphibians have different life stages including egg, embryo, tadpole, juvenile, and adult for an aquatic phase and a terrestrial phase, it is important to understand chemical toxicity for testing in different life stages as well as considerable economic and experimental efforts. There are some limitations to the availability of experimental toxicity data in amphibians and the currently limited requirements to address such taxa as ecotoxicological targets. Thus, the use of new approach methodologies (NAMs) has been applied to nontarget species such as honey bees and collembola [271, 272]. NAMs include QSAR/QSTR models, which provide an effective way to predict chemical and toxicological properties based on the structural properties of the chemical [273]. QSAR/QSTR models for tadpoles may be particularly pertinent because they are fully aquatic, and may be exposed to a range of chemicals throughout their developmental stages making them potentially sensitive as they undergo metamorphosis. Due to the limitation of experimental data, insufficient QSAR/QSTR models have been published for frog tadpoles including the prediction of acute toxicity for benzene derivatives in *Rana japonica*, a limited number of alcohol compounds in *Rana temporaria*, *Rana chensinensis*, and for undescribed species [209, 274–279].

Toropov et al. [280] developed a regression-based QSAR/QSTR model to predict the acute toxicity of aromatic chemicals in tadpoles of the Japanese brown frog (*R. japonica*) using available acute toxicity data and the CORAL software (http://www.insilico.eu/coral) based on the Monte Carlo optimization with the so-called index of the ideality of correlation [281].

Toropov et al. [280] introduced five random splits via the index of ideality of correlation (IIC) optimization from the Monte Carlo optimization carried out with the so-called IIC, that is, a special component of the target function described in the literature [281]. They obtained QSAR models for the prediction of acute toxicity in tadpoles of *R. japonica*. For their work, the 12 h time point was selected because it was the most common measurement in *R. japonica* tadpoles. They used the structures associated with the chemicals being modeled in work using the CORAL models, which were represented by the simplified molecular input line entry system (SMILES) [282]. The CORAL model is the one-variable correlation between the SMILES-based 2D descriptor and the acute toxicity endpoint ($-\log LC_{50}$), where LC_{50} is the acute median lethal molar concentration for 50% of *R. japonica* tadpoles. Toropov et al. [280] obtained the third random split via the IIC-optimization QSAR/QSTR model for the prediction of acute toxicity in tadpoles of the Japanese brown frog (*R. japonica*) ($-\log LC_{50}$) as follows:

$$-\log LC_{50}(Rana\ japonica, M, 12\ h) = 1.9027 + 0.7199 \sum CW(S_k) \qquad (4.6^*)$$

where S_k is a SMILES atom, that is, one symbol (e.g., "C," "c," "N," and "O") or a group of symbols that cannot be examined separately (e.g., "Cl" and "Br"). $CW(S_k)$ provides the correlation weight of the S_k corresponding to a coefficient that is combined with the value of the descriptor if the corresponding SMILES contains S_k.

It is possible to convert the name or chemical structure of a chemical to SMILES through software or an online website, for example, http://cdb.ics.uci.edu/cgibin/Smi2 DepictWeb.py. Thus, the use of eq. (4.6*) requires SMILES and correlation weights $(CW(S_k))$ obtained by Monte Carlo optimization, which is given in Table 4.2.

Table 4.1: Correlation weights $(CW(S_k))$ obtained by Monte Carlo optimization for eq. (4.6*).

S_k	$CW(S_k)$	S_k	$CW(S_k)$
(.	−0.0813	Cl.	1.0972
1.	0.8421	N.	0.0
2.	1.0824	O.	0.1401
=.	−0.2139	S.	0.0
C.	0.4530	[N+].	1.1961
F.	−0.2871	[O−].	0.1351
Br.	1.1052	c.	−0.1438

Example 4.1: 2-Bromo-4-methyl phenol has the following molecular structure:

(a) Calculate $\sum CW(S_k)$ from SMILES. (a) Use eq. (4.6*) to calculate −log LC_{50}. (b) If the experimental value of −log LC_{50} is 3.7200, calculate the deviation of eq. (4.6*).

Answer: (a) The use of the website http://cdb.ics.uci.edu/cgibin/Smi2DepictWeb.py gives SMILES as Cc1cc (Br)c(O)cc1. Thus, the value of $\sum CW(S_k)$ using Table 4.1 gives $\sum CW(S_k) = 12(-0.1438) + 2(0.1401) + 2(0.8421) + 2(-0.0813) + 2(1.0824) = 2.1945$

(b) −log LC_{50}(Rana japonica, M, 12 h) $= 1.9027 + 0.7199 \sum CW(S_k)$
$$= 1.9027 + 0.7199(2.1945) = 3.4825$$

(c) Dev = 3.7200−3.4828 = 0.2375

4.1.3 Chemometric Modeling of Acute Toxicity of Diverse Aromatic Compounds Against *Rana japonica*

Nath and Roy [265] developed five QSAR/QSTR models to predict the acute toxicity of different aromatic compounds against the tadpoles of *R. japonica* using the LC_{50} database of Toropov et al. [280], which were described in Section 4.1.2. They showed some important structural features and molecular properties in the form of descriptors in the following models:

Model IM 1

$$-\log LC_{50}(Rana\,japonica, M, 12\,h) = 0.14166 - 0.45177S3K + 1.40881LOC + 0.39671B01[C-O]$$
$$+ 0.00172TPSA(NO) + 1.05042\log K_{\text{OW,WC}}$$

$$(4.7^*)$$

Model IM 2

$$-\log LC_{50}(Rana\,japonica, M, 12\,h) = 0.1577 - 0.44164S3K + 1.43666LOC + 0.39648B01[C-O]$$
$$+ 0.00439T(O\ldots O) + 1.04042\log K_{\text{OW,WC}}$$

$$(4.8^*)$$

Model IM 3

$$-\log LC_{50}(Rana\,japonica, M, 12\,h) = -0.27714 - 0.43625S3K + 1.61505LOC + 0.40285B01[C-O]$$
$$+ 3.62864PW5 + 1.04125\log K_{\text{OW,WC}}$$

$$(4.9^*)$$

Model IM 4

$$-\log LC_{50}(Rana\,japonica, M, 12\,h) = 0.1517 - 0.43534S3K + 1.44357LOC + 0.40751B01[C-O]$$
$$+ 0.00892N\% + 1.03376\log K_{\text{OW,WC}}$$

$$(4.10^*)$$

Model PDM

$$-\log LC_{50}(Rana\,japonica, M, 12\,h) = 0.15883 - 0.42458S3K + 1.39579LOC + 0.4432B01[C-O]$$
$$+ 0.01451N\% - 0.00025TPSA(NO) + 1.03232\log K_{\text{OW,WC}}$$

$$(4.11^*)$$

Among different descriptors given in eqs. (4.7*) and (4.8*), two descriptors $TPSA(NO)$ and $\log K_{\text{OW,WC}}$ were described in Section 4.1.1.2. Since the coefficient of $\log K_{\text{OW,WC}}$ has a high value, it contributes the high impact toward $-\log LC_{50}(M, 12\,h)$.

$B01[C-O]$ is a 2D atom pair-type descriptor, which has also a high contributing descriptor toward C–O. It identifies the presence of carbon and oxygen atoms at the topological distance one in which 1 and 0 suggest the presence and absence, respectively, of the C and O at the topological distance 1 [283].

$S3K$ is a topological descriptor, which is 3 paths Kier alpha-modified shape index. Kier alpha-modified shape descriptor defines different shape contributions of heteroatoms and hybridization states of a molecule, which is a function of the number of atoms and their bonding relationship in a molecule [284]. Since the coefficients of S3K are negative, toxicity will be higher for molecules with a low S3K value.

$T(O\ldots O)$ is a 2D atom pair index, which is the sum of topological distances between two oxygen atoms in a molecule. The positive coefficients of $T(O\ldots O)$ in the equations suggest that the toxicity of the investigated compounds increases with the higher values of $T(O. . .O)$.

$PW5$ is a class of topological index, which is path/walk 5 – Randic shape index. Since the coefficient of $PW5$ in eq. (**4.9***) has a positive sign, the toxicity of investigated compounds increases with the increasing values of $PW5$.

$N\%$ is a constitutional index, which shows the % of nitrogen atoms present in a molecule as follows:

$$N\% = \frac{\text{Total number of nitrogen atoms in a molecule}}{\text{Total number of atoms in a molecule}} \times 100 \qquad (4.12)$$

Since nitrogenous compounds can decompose to ammonia, nitrite, or nitrate, they are considered toxic to aquatic species. Thus, they promote biochemical and histological changes and decrease the oxygen-carrying capacity in the blood and affect different physiological functions [285, 286]. The positive coefficients of $N\%$ in eqs. (**4.10***) and (**4.11***) suggest that the toxicity of investigated compounds increases with increasing $N\%$.

LOC is a topological descriptor, which is a lopping-centric index. It is an index defined as the mean information content derived from the pruning partition of the structure molecular graph or acyclic graph. It describes the number of terminal vertices removed at the kth steps (to remove all graph vertices) [287]. It belongs to "centric indices" that have been projected for the quantification of the degree of compactness of a molecule by distinguishing between molecular structures arranged differently concerning their centers [287]. It is linked with the surface and symmetry of the two-dimensional representation of a molecule [288]. Due to its positive coefficients in eqs. (**4.7***)–(**4.11***), an increase in the LOC values causes a significant increase in toxicity.

4.1.4 Toxicity of Different Substituted Aromatic Compounds to the Aquatic Ciliate Tetrahymena pyriformis

Luan et al. [289] developed several QSAR models using the MLR method to identify and predict the acute toxicity (IC_{50}) of substituted aromatic compounds to the aquatic ciliate *T. pyriformis*. They divided the whole data set into three groups concerning the important function group of the substituted aromatic compounds and introduced three correlations as follows:

Group 1: Compounds with nitro groups

$$-\log IC_{50}(mg\ L^{-1}) = 73.123 + 0.002G^2 - 1.016P_O - 1.082E_{nn}(C - H) \qquad (4.13^*)$$

G^2 refers to gravitation indexes for all bonded pairs of atoms, which is defined as

$$G^2 = \sum_{i>j}^{N_b} \frac{m_i m_j}{r_{ij}^2} \tag{4.14}$$

where m_i and m_j are the atomic weights of atoms i and j; r_{ij} is the interatomic distance; N_b is the number of bonds in the molecule.

P_O belongs to the valency-related descriptors, which relate to the strength of intermolecular bonding interactions. It characterizes the stability of the molecules, their conformational flexibility, and other valency-related properties [290].

$E_{nn}(C–H)$ is the maximum nuclear–nuclear repulsion energy for a C–H bond. It is calculated as follows:

$$E_{nn}(\text{CH}) = \frac{Z_C Z_H}{R_{CH}} \tag{4.15}$$

where Z_C and Z_H are the nuclear (core) charges of atoms C and H, respectively; R_{CH} is the distance between them. This energy defines the nuclear repulsion-driven processes in the molecule, which may be related to the conformational (rotational and inversional) changes or atomic reactivity in the molecule [291].

Group 2: Compounds with halogen substituents

$$-\log IC_{50}(mg\ L^{-1}) = -18.030 + 0.438 \log P - 6.605 PNSA\_2/TMSA + 17.066 P_{\text{SIGMA}} \quad \textbf{(4.16*)}$$

log P stands for the solvation characteristics (hydrophobicity of chemicals). It is closely related to the change in the Gibbs energy of the solvation of a solute between two solvents.

$PNSA\_2/TMSA$ is $FNSA\_2$ fractional $PNSA$ ($PNSA\_2/TMSA$) [Zefirov's PC]. It contributes to the calculation of atomic partial charges to the total molecular solvent-accessible surface area [291].

P_{SIGMA} represents the maximum bond order for a given pair of atomic species in the molecule. It has values for a given pair of atomic species in the molecule with the lower limit P_{SIGMA} (min) > 0.1.

Group 3: Compounds containing both nitro and halogen substituents

$$-\log IC_{50}(mg\ L^{-1}) = -16.640 - 2.141 I_C + 0.151 E_{nn}(C – C) - 51.290 RPCG \quad \textbf{(4.17*)}$$

I_C is a geometrical descriptor that relates to the atomic masses and the distance of the atomic nucleus from the main rotational axes. It characterizes the mass distribution in the molecule.

$E_{nn}(C–C)$ is the maximum nuclear–nuclear repulsion energy for a C–C bond. It is calculated as follows:

$$E_{nn}(CC) = \frac{Z_C Z_C}{R_{CC}} \tag{4.18}$$

where Z_C and Z_C are the nuclear (core) charges of C atoms; R_{C-C} is the distance between them. This energy illustrates the nuclear expulsion-driven processes in the molecule, which may be related to the conformational (rotational and inversional) changes or atomic reactivity in the molecule [292].

RPCG is a relative positive charge, which belongs to electrostatic descriptors. Due to the negative sign of its coefficient, the relative positive charge of the molecule is negatively related to the endpoint values ($-\log IC_{50}$).

In summary, the repulsion between the two bonds and the local charge on the surface of the molecule appeared in different models. Thus, these two factors have a greater influence on the structure of the compound and should be relatively valued.

4.1.5 Toxicity of Aromatic Pollutants and Photooxidative Intermediates in Water

Since many aromatics are nonbiodegradable, they provide difficulties in their removal by conventional wastewater treatment plants (WWTPs), based on physical and biological processes as primary and secondary treatment methods, respectively [293]. Thus, conventional WWTPs require a tertiary treatment based on advanced methods for water purification such as the advanced oxidation process (AOP). This treatment generates highly reactive and unselective species such as hydroxyl radicals (HO•) [294]. A formation of more toxic and mutagenic products may occur at certain stages of AOP treatment of aromatics.

Cvetnic et al. [295] developed QSAR/QSTR models for predicting the toxicity toward *Vibrio fischeri* of aromatics and their reaction mixtures during the treatment by AOPs. They selected the set of single-benzene ring compounds (S-BRC) diversified by the type and number of substituents. They determined the toxicity of parent S-BRCs in terms of EC_{50} (effective concentration causing 50% reduction of bioluminescence) and toxicity units (TU_0). They also determined the half-life toxicity ($TU_{1/2}$) by treating the S-BRCs with a photooxidative AOP: UV-C/H_2O_2. They correlated experimentally determined toxicity parameters with structural features of studied S-BRCs over molecular descriptors derived from their optimized structures using QSAR/QSTR approaches as follows [295]:

$$-\log EC_{50} = -3.390 + 0.504 GATS6e + 0.341 GATS7s + 2.688 SPH - 1.453 R1e \tag{4.19*}$$

$$-\log TU_0 = -1.510 + 0.798 MATS6m + 0.255 MATS7s + 0.857 Ve + 0.835 HATS1v \tag{4.20*}$$

$$-\log TU_{1/2} = -2.357 - 0.413 Psi\_i\_1d + 2.182 ATSC3i - 0.881 Mor18s$$
$$+ 0.789 Mor22s - 1.205 HATS0e \tag{4.21*}$$

Equation (**4.19***) contains two 2D autocorrelations (*GATS6e* and *GATS7s*), one geometrical (*SPH*), and one GETAWAY (GEometry, Topology, and Atom-Weights AssemblY) descriptor (*R1e*). As shown in eq. (**4.19***) and from the values of coefficients of descriptors, *SPH* has the highest contribution to the endpoint while *GATS6e* and *GATS7s* descriptors contribute much less to toxicity prediction. Among the four descriptors, only R1e has a synergistic contribution to the endpoints because, by its increment, the toxicity of compounds decreases. Meanwhile, *SPH*, *GATS6e*, and *GATS7s* descriptors contribute to the toxicity in an antagonistic way because their increments decrease the value of EC_{50} corresponding to toxicity increases. Since *SPH* pertains to geometrical descriptors, most descriptors from this category are calculated directly from the 3D representation of the molecule counting the molecule atoms and other quantities from *x*, *y*, and *z* coordinates, which include numerous information on molecule's geometry, size, shape, and topology. Moreover, *SPH* measures the molecule flatness. It is important to consider the shape of a molecule because it dictates the activity of various microorganisms [296]. 3D complementarity is of great importance when modeling the activity involving the interaction of the lock-and-key type such as toxicity [297]. The number, type, and position of simple substituents in the congeneric series of S-BRCs influence toxicity. The *R1e* descriptor pertains to a group of GETAWAY descriptors, which are derived from the molecular influence matrix (MIM) corresponding to represent the molecular structure through the development of particular output [298]. There are two groups of GETAWAY descriptors: (1) descriptors computed using traditional matrix operators and information theory concepts onto MIM and influence/distance matrix R; (2) descriptors derived using the spatial autocorrelation formulas that weight the molecule atoms simultaneously accounting for properties such as polarizability, atomic mass, van der Waals volume, and electronegativity together with 3D information encoded by the elements of MIM and R. The second group of GETAWAY descriptors may also be weighted with some property associated with the atoms such as *p* (polarizability), *m* (relative atomic mass), *e* (Sanderson electronegativity), *v* (van der Waals volume), *i* (ionization potential), and *s* (E-state; electrotopological states) [298]. Since R1e reflects a weighting scheme over electronegativity, its importance is confirmed for toxicity, for example, Melagraki et al. [299] showed the relationship between equalized electronegativity and toxicity of organic compounds toward *V. fischeri*. Two descriptors *GATS6e* and *GATS7s* pertain to 2D autocorrelations, which provide more correctly its subgroup Geary (GATS) autocorrelations. Two further 2D autocorrelations include Broto–Moreau, also known as autocorrelation of a topological structure (ATS), and Moran (MATS) descriptors. These descriptors are computed using molecular graphs, which are classified by the term "the lag" with a pertaining number and particular weighting scheme (*m*, *p*, *e*, *v*, *i*, and *s*) [298]. The products of atomic weights of the terminal atoms of all paths of the considered path length can be summed to calculate the "lags." *GATS6e* and *GATS7s* are weighted by electronegativity and atomic electrotopological states, respectively. Electrotopological state atom (E-state) indices indicate the important topological features and molecular fragments mediating a particular response [300]. It was shown that the E-state indices are important in toxicity modeling [301].

Since there is a straightforward relationship between EC_{50} values and TU_0 values for S-BRC solutions with an initial concentration of 1 mM, they are connected over molecular mass. Equation (**4.20***) also contains two 2D autocorrelation descriptors, which include the Moran subgroup (*MATS6m* and *MATS7s*). *MATS6m* and *MATS7s* possess identical lag numbers as those Geary ones included in eq. (**4.19***). *MATS6m* is weighted by atomic mass, which provides the missing link between the models. Dimensionless TU_0 is correlated with EC_{50} over the molecular mass of S-BRCs. MATS7s has even the same weighing scheme (E-state) as its Geary analog in eq. (**4.19***). Equation (**4.20***) also includes WHIM (*Ve*) and GETAWAY (*HATS1v*) descriptors. WHIM descriptors are based on statistical indices, which can be calculated on the projections of atoms along the principal axes [298]. They capture relevant molecular 3D information regarding the molecular shape, size, symmetry, and atom distribution as compared to invariant reference frames. They can be also weighted by m, p, e, v, i, and s, which were described for GETAWAY and 2D autocorrelation descriptors. The WHIM descriptor shows V total size index weighted by Sanderson electronegativity. Therefore, electronegativity is already depicted as an important structural characteristic of S-BRCs in predicting toxicity toward *V. fischeri*. The determined volume with parameters such as the shape and size of the molecule plays an important role in such mechanisms.

The descriptors of eq. (**4.20***) pertain to several classes: topological indices (*Psi_i_1d*), 2D autocorrelations (*ATSC3i*), 3D Molecule Representation of Structures based on Electron diffraction (3D-MoRSE: *Mor18s* and *Mor22s*), and GETAWAY (*HATS0e*). Toxicity in half-life is possessed by both parent SBRCs in the remaining concentration (50% of the initial value) and formed by-products upon HO• attack. The highest contribution in eq. (**4.20***) is obtained by the *ATSC3i* descriptor, which is calculated using the Broto–Moreau algorithm and weighted by ionization potential. The main degradation mechanism by HO•-driven processes involves H-abstraction (at either C-ring or C-substituent atom) and subsequent hydroxylation [302]. The polyhydroxybenzene (PHB) derivatives are obtained in the first step, which then undergoes the ring-opening reactions, and eventually, such aliphatics are mineralized. Since aromatics still strongly prevail over aliphatics, the reaction mixture is consistent with parent pollutants and PHBs. PHBs are known for their toxicity toward *V. fischeri* [303]. $TU_{1/2}$ are in most cases higher than TU_0 values because by the proposed pathways of studied S-BRCs, most likely various PHBs indicate the formation of toxic by-products [304]. The involvement of ionization potential over the weighting scheme in *ATSC3i* can also be related to the degradation kinetics of S-BRCs studied. Ionization potential can be considered as an absolute value of E_{HOMO}, which is highly correlated with the reactivity of organics toward radical attack [305]. Equation (**4.20***) includes also GETAWAY (*HATS0e*) and 3D-MoRSE descriptors (*Mor18s* and *Mor22s*), which are weighted with already pointed out significant molecular properties influencing toxicity. Electronegativity and E-state provide a clear relationship to the structural features of S-BRCs, which contribute to the toxicity of S-BRCs and their degradation intermediates. 3D-MoRSE class include descriptors that are computed by summing atom weights viewed by different angular scattering function. They are tagged as Mor*sw* where *s* is in the range from 1 to 32, while w denotes a particular weighting scheme (m, p, e, v, i, and s) [298]. $Mor18s$ and $Mor22s$ descriptors are weighted by the same scheme (E-state) but descriptors possess opposite contributions to the endpoint according to the coefficients of these variables in eq. (**4.20*****). Psi_i_1d descriptor is intrinsic state pseudoconnectivity index – type 1d, which has the lowest contribution to the endpoint. It pertains to the new class of topological descriptors calculated by DRAGON software. Intrinsic state molecular pseudo connectivity indices are based on the intrinsic state concept that is built on the intrinsic and the electrotopological state values [306]. They are useful in the prediction of biodegradability and toxicity of some chemical classes [307].

In brief, the toxicity of parent S-BRCs depends on their electronegativity, geometry (size, shape, and volume of molecule), and E-state, combining electronic and topological characteristics of atoms/molecules. Meanwhile, the toxicity of S-BRC reaction mixtures in half-life depends on ionization potential, electronegativity, and E-state, as well as their topological characteristics.

4.1.6 Risk Assessment of Aromatic Compounds to *Tetrahymena pyriformis* by a Simple QSAR/QSTR Model

A simple QSAR/QSTR model is introduced for reliable prediction of the toxicity of organic aromatic compounds based on the endpoint $-\log IC_{50}(mM)$ toward *T. pyriformis*. It uses an experimental data set of $-\log IC_{50}(mM)$ for 892 organic aromatic compounds. It uses additive variables as core correlation, which includes the number of nitro groups, carbon and halogen atoms, as well as some specific polar groups and molecular weight. It also uses two nonadditive correcting functions for the increment of the reliability of the core correlation as follows:

$$-\log IC_{50}(mM) = -1.694 + 0.1649n_C + 0.3124n_{Hal} + 4.772 \times 10^{-3}MW + 0.3799n_{NO_2}$$

$$-0.1796n_{OR+1.5CN+CO} - 0.5378n_{NHR+NR_2} + 0.7179(IC_{50})^+ - 0.6263(IC_{50})^-$$

$$(4.22^*)$$

where n_C, and n_{Hal} are the number of carbon and halogen atoms, respectively; MW is the molecular weight; n_{NO_2} is the number of nitro groups; $n_{OR+1.5CN+CO}$ is the sum of ether, cyano (multiplied by 1.5), and carbonyl groups; n_{NHR+NR_2} is the sum of primary and secondary amine groups; $(IC_{50})^+$ and $(IGC_{50})^-$ are positive and negative contributions of some specific molecular fragments and isomers, respectively. Table 4.2 provides the optimized values of $(IC_{50})^+$ and $(IGC_{50})^-$.

Table 4.2: The optimized values of $(IC_{50})^+$ and $(IGC_{50})^-$.

No.	Structural isomers	Condition	$(IGC_{50}^{-1})^+$	$(IGC_{50}^{-1})^-$	Example
1			0	2.0	
2	Ar-SO$_2$-OH or -OR		0	2.0	
3			0	2.0	
4	Ar-O-(CH$_2$)$_{n=1}$ or $_2$-OH or -CONH$_2$		0	2.0	

(continued)

Table 4.2 (continued)

No.	Structural isomers	Condition	$(IGC_{50}{}^{-1})^+$	$(IGC_{50}{}^{-1})^-$	Example
5			0	2.0	
6			0	2.0	
7	Ar-CHR-(OH or -NH$_2$)	R is a hydrogen atom or alkyl group and the aromatic ring does not contain two nitro groups	0	1.5	
8	Ortho F$_2$-Ar-CHO	Two fluorine atoms are ortho to the aldehyde group	0	1.5	
9		Ortho	0	2.0	
			0	1.0	
10		R' containing CO group	0	1.0	

#	Name / Condition	Meta	Ortho or para
11		0	1.0
12	Ortho or para	1.0	0
13	Ortho HO-Ar-OR' — R' does not contain $-(CH_2)_{n=1\ or\ 2}-OH$ or NH_2	0	1.0
14	Ortho HO-Ar-OCOR' — R' is aryl or alkyl group containing –OH group	0	1.0
15	Ar-trifluoromethyl — Without ortho $-NO_2$ to trifluoromethyl	0	1.0
16	Ar-X(R_1)(R_2)(-OH or -NH_2) — X = alkyl group with more than one carbon atoms; R_1 and R_2 = hydrogen atoms or an alkyl group	0	1.0
17	R_1 and R_2 = saturated or an unsaturated alkyl group	0	1.0
18	Dioate ester	0	1.0

(continued)

Table 4.2 (continued)

No.	Structural isomers	Condition	$(IGC_{50}^{-1})^{+}$	$(IGC_{50}^{-1})^{-}$	Example
19	Benzene ring containing both one -NO$_2$ and one or two fluorine groups	Without the attachment of the polar group to the aromatic ring	0	1.0	
20	or Ar-SH or Ar-C≡C-CHO		2.0	0	
21	Benzene ring containing two -NH$_2$ (or –OH) in the para position	Containing one or two methyl groups	2.0	0	
		Containing three or four methyl groups or one bulky hydrocarbon group or one aromatic group	1.0	0	
		Containing one halogen atom	1.5	0	
		Containing two chlorine atoms with a plane of symmetry	1.5	0	

No.	Description	Condition	Value	Structure	
22	Benzene ring containing -NH$_2$ (or -OH) and -NO$_2$ in para position as well as further fluorine atom or one (or two) methyl groups		2.0		0
23	Benzene ring containing one -CHO and two -OH groups	Three substituents are in the ortho position	1.0		0
24	Ar-N = C = S or -NHC(=S)NH$_2$	-N = C = S group may be directly attached to an aromatic ring or through -(CH$_2$)$_n$- (or -CH<) group	No. of NCS or NHCSNH$_2$ × 2		0
25	Benzene ring containing three -OH groups	At least two -OH are ortho to each other	1.0		0
26	or aromatic compound containing 1,4-oxathiane ring		1.0		0
27		R is an alkyl group with more than five carbon atoms	1.0		0
28	Benzene ring containing one -NH$_2$ (or -OH) as well as more than one chlorine atoms	Except for -OH between two chlorine atoms	1.0		0
29		X = Deactivating groups (e.g., nitro and halogen groups)	1.0		0

(continued)

Table 4.2 (continued)

No.	Structural isomers	Condition	$(IGC_{50}{}^{-1})^+$	$(IGC_{50}{}^{-1})^-$	Example
30		X and Y are halogen atoms	1.0	0	
31		R = Linear alkyl group with more than three carbon numbers or containing –CH=C(Cl)–CHO or including –C≡C–C(O)O–	1.0	0	
32	Benzene ring containing –OH, -Cl (or –Br), and –CHO	Except –OH ortho to –Br atom	1.0	0	
33	Benzene ring containing –NO₂ -Cl (or –Br) and –CN		1.0	0	

Example 4.2: 3,5-Dichloro-2-hydroxybenzaldehyde has the following molecular structure:

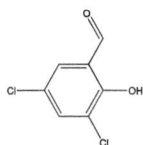

Chemical Formula: $C_7H_4Cl_2O_2$
Molecular Weight: 191.01

(a) Use eq. (**4.22*****) to calculate $-\log IC_{50}(mM)$. (b) If the experimental value of $-\log IC_{50}(mM)$ is 1.55, calculate the deviation of eq. (**4.22*****). (c) Su et al. [308] used support vector regression with 68 molecular descriptors for prediction of $-\log IC_{50}(mM)$ of aromatic compounds. If the predicted result of their method for this compound is 1.03, which method gives closer output as compared to the measured value?

Answer: (a) The use of eq. (**4.22*****) and Table 4.1 gives

$$-\log IC_{50}(mM) = -1.694 + 0.1649n_C + 0.3124n_{Hal} + 4.772 \times 10^{-3} MW + 0.3799n_{NO_2}$$
$$- 0.1796n_{OR+1.5CN+CO} - 0.5378n_{NHR+NR_2} + 0.7179(IC_{50})^+ - 0.6263(IC_{50})^-$$

$$-\log IC_{50}(mM) = -1.694 + 0.1649(7) + 0.3124(2) + 4.772 \times 10^{-3}(191.01) + 0.3799(0)$$
$$- 0.1796(1) - 0.5378(0) + 0.7179(1.0) - 0.6263(0) = 1.54$$

(b) Dev = 1.53–1.54 = –0.01
(c) Since the deviation of the method by Su et al. [308] is –0.52, eq. (**4.22*****) gives a closer result.

4.1.7 Toxicity Toward *Chlorella vulgaris* of Organic Aromatic Compounds in Environmental Protection

IC_{50} and IC_{20} (concentration producing 20% inhibition) may measure the potency of chemicals to algae with batch assays using different durations (i.e., 48, 72, and 96 h) [309], which can provide different endpoints (e.g., acute, chronic, and low toxic effects) from these assays [310]. The low toxic effect endpoints can be considered chronic values because several generations are produced during these assays [311]. *NOEC* shows the compound concentration that causes no significant difference but *LOEC* gives the lowest concentration that makes a significant difference compared to controls [312].

There are some correlations between low-level toxicities and complex descriptors. Austin and Eadsforth [313] developed QSAR/QSTR models for *NOEC* values of different nonpolar narcotic chemicals using a fish toxicity data set. Tugcu and Sacan [314] used both 2D and 3D descriptors as well as $-\log IC_{50}$ for modeling $-\log NOEC$ and $-\log IC_{20}$ endpoints. Application of their models is limited to the availability of $-\log IC_{50}$ values for the demanded compounds. The use of 3D descriptors increases computational complexity and restrictions of the easy transferability of their models. For chronic toxicity forecast [315], Seth and Roy [316] improved QSAR/QSTR models to assess low-level algal toxicities. They employed PLS regression using low-level toxicity values against algal species as well as two 2D descriptors containing ETA and non-ETA indices. Their models can be used for four endpoints $-\log IC_{50}$, $-\log IC_{20}$, $-\log LOEC$, and $-\log NOEC$ of phenol and aniline derivatives toward *Chlorella vulgaris* (*C. vulgaris*). Yan et al. [317] proposed QSAR/QSTR models with the same mathematical structure for predicting $-\log IC_{50}$, $-\log IC_{20}$, $-\log LOEC$, and $-\log NOEC$ toward *C. vulgaris* of substituted phenols and anilines based on norm index descriptors. These approaches need complex descriptors, computer codes, and expert users. Suitable structural descriptors rather than complex descriptors for finding endpoints $-\log IC_{50}$, $-\log IC_{20}$, $-\log LOEC$, and $-\log NOEC$ toward *C. vulgaris* of aromatics can be used as follows:

$$-\log IC_{50}(mM) = -0.9144 + 0.3185n_{Cl} + 0.6021C_{NO_2} + 0.0071MW$$
$$+ 0.7570C_{Intra\,H-bond} + 1.0890IISP - 1.2901DISP \tag{4.23*}$$

$$-\log IC_{20}(mM) = -0.9955 + 0.2401n_{Cl} + 0.5052C_{NO_2} + 0.0102MW$$
$$+ 0.6979C_{Intra\,H-bond} + 0.7408IISP - 1.1870DISP \tag{4.24*}$$

$$-\log LOEC(mM) = -0.8605 + 0.2542n_{Cl} + 0.4982C_{NO_2} + 0.0107MW$$
$$+ 0.5015C_{Intra\,H-bond} + 0.4800IISP - 1.4902DISP \tag{4.25*}$$

$$-\log NOEC(mM) = 0.3846n_{Cl} + 0.5312C_{NO_2} + 0.0065MW +$$
$$0.6155C_{Intra\,H-bond} + 0.4125IISP - 1.5652DISP \tag{4.26*}$$

where n_{Cl} and MW are the numbers of chlorine atoms and molecular weight, respectively. Two descriptors C_{NO_2} and $C_{Intra\,H-bond}$ are the contribution of nitro groups and intramolecular hydrogen bonding to a hydroxyl group, respectively, which are described as follows:

(a) C_{NO_2}: The value of C_{NO_2} equals 1.0 for the presence of NO_2 group in phenol derivatives except nitro group between two substituents without alkyl groups and more than one chlorine atom for mononitro derivatives. The value of C_{NO_2} is also 1.0 for the existence of more than one nitro group in aniline derivatives if two nitro groups are in the ortho-position or one nitro group without further substituent in the ortho-position of the amino group.

(b) $C_{Intra\,H-bond}$: The values of $C_{Intra\,H-bond}$ equal 1.3, 0.3, and 0.5 for the presence of $-$OR ortho to $-$OH in the presence of more than one $-$OH group, a halogen atom except for more than two halogen atoms where $-$OH group exists between two halogen atoms, and NO_2 groups, respectively, as well as 0.6 and 1.5 for the existence of two and three adjacent $-$OH groups, respectively.

For two descriptors *IISP* and *DISP*, Table 4.3 provides the molecular structure of some classes of aromatic compounds corresponding to the best values of these descriptors.

Nitroaromatic energetic compounds are widely used as explosives where they can show several manifestations of toxicity in humans including skin sensitization, immunotoxicity, and methemoglobinemia [318]. Equations (**4.23***)–(**4.26***) can be used for

Table 4.3: The optimized values of *IISP* and *DISP*.

No.	Structural isomers	Condition	IISP	Example
1	Benzene ring containing –OH groups as well as chlorine and methyl groups or aza group	Except *ortho*-chlorine group to –OH	0.7	
2	Mononitrophenol	Except *ortho*-nitro group to –OH	0.6	
3	Aniline containing more than one nitro group	Except *ortho*-nitro group to –NH$_2$	1.0	
4	Phenol or phenol containing more than one bromine or chlorine atoms	Except *ortho*-chlorine group to –OH	0.7	
5	Benzene ring containing only –OH groups	Except for *ortho*- or *para*-hydroxyl group to -OH	0.7	
6	Phenol containing more than one alkyl group	Except *ortho*-alkyl group to –OH	0.5	
		Including *ortho*-alkyl group to –OH	0.4	
7	The presence of –CHO with an alkyl group containing more than two carbon atoms or chlorine atom or –NRR′ group	Except for *ortho*-chlorine atom to –CHO	0.7	
8	The existence of the –NO$_2$ group with one of the following groups: (i) Ortho nitro groups (ii) One chlorine atom, which may be in the presence of alkyl groups (iii) P(S)(O)$_3$ and alkyl groups attached to the benzene ring	–	0.7	

Table 4.3 (continued)

No.	Structural isomers	Condition	IISP	Example
9	Benzene ring containing –OH groups and one methyl group or the presence of both bromine atoms and nitro groups simultaneously or only alkoxy group	Except for *ortho*-methyl group to -OH	0.5	
10	Aniline or aniline containing alkyl groups	–	1.0	
11	Aniline containing halogens or hydroxyl groups	–	0.5	
12	The presence of one –CHO or one –NO$_2$ or only alkoxy group without the other substituents	Without the presence of polar groups or an alkyl group containing more than one carbon atom	0.5	
13	The presence of –CON< or -COOR	–	0.7	
14	Benzene containing one (or two) –OH and more than three chlorine atoms where two chlorine atoms are in ortho position to –OH or the presence of fluorine atom or alkoxy group or –CHO group	–	0.7	

some of the new and old explosives since their –log IC_{50}, –log IC_{20}, –log $LOEC$, and –log $NOEC$ toward *C. vulgaris*. Table 4.4 shows the predicted values of –log IC_{50}, –log IC_{20}, –log $LOEC$, and –log $NOEC$ toward *C. vulgaris* for several important nitro-aromatic explosives including ICM-102, TATB, TNT, HNS, and LLM-105. Among the mentioned explosives, HNS is a more potent compound as compared to the other explosives because it contains two aromatic rings and provides high values of –log IC_{50}, –log IC_{20}, –log $LOEC$, and –log $NOEC$ toward *C. vulgaris*.

Table 4.4: The calculated values of $-\log IC_{50}$, $-\log IC_{20}$, $-\log LOEC$, and $-\log NOEC$ toward *C. vulgaris* by **(4.23*)**–**(4.26*)** for several explosives.

Name	Structure	$-\log IC_{50}$	$-\log IC_{20}$	$-\log LOEC$	$-\log NOEC$
ICM-102		0.5282	1.0585	1.2948	1.3139
TATB		1.5300	2.1327	2.3901	2.2091
TNT		0.7065	1.3123	1.5611	1.4763
HNS		2.2985	3.5790	3.9397	2.9263
LLM-105		0.6279	1.2005	1.4438	1.4047

Example 4.3: Using eqs. **(4.23*)**–**(4.26*)**, calculate the values of $-\log IC_{50}$, $-\log IC_{20}$, $-\log LOEC$, and $-\log NOEC$ toward *C. vulgaris* for 4-nitrophenol with the following molecular structure:

Chemical Formula: $C_6H_5NO_3$
Molecular Weight: 139.11

Answer: (a) The use of eqs. **(4.23*)**–**(4.26*)** and Table 4.4 (no. 2) gives

$$-\log IC_{50}(mM) = -0.9144 + 0.3185n_{Cl} + 0.6021C_{NO_2} + 0.0071MW$$
$$+ 0.7570C_{\text{Intra H-bond}} + 1.0890IISP - 1.2901DISP$$
$$-\log IC_{50}(mM) = -0.9144 + 0.3185(0) + 0.6021(1) + 0.0071(139.11) + 0.7570(0)$$
$$+ 1.0890(0.6) - 1.2901(0) = 1.334$$
(4.23*)

$$-\log IC_{20}(mM) = -0.9955 + 0.2401n_{Cl} + 0.5052C_{NO_2} + 0.0102MW$$
$$+ 0.6979C_{\text{Intra H-bond}} + 0.7408IISP - 1.1870DISP$$
$$-\log IC_{20}(mM) = -0.9955 + 0.2401(0) + 0.5052(1) + 0.0102(139.11)$$
$$+ 0.6979(0) + 0.7408(0.6) - 1.1870(0) = 1.368$$
(4.24*)

$$-\log LOEC(mM) = -0.8605 + 0.2542n_{Cl} + 0.4982C_{NO_2} + 0.0107MW$$
$$+ 0.5015C_{\text{Intra H-bond}} + 0.4800IISP - 1.4902DISP$$
$$-\log LOEC(mM) = -0.8605 + 0.2542(0) + 0.4982\ (1) + 0.0107(139.11)$$
$$+ 0.5015(0) + 0.4800(0.6) - 1.4902(0) = 1.409$$
(4.25*)

$$-\log NOEC(mM) = 0.3846n_{Cl} + 0.5312C_{NO_2} + 0.0065MW$$
$$+ 0.6155C_{\text{Intra H-bond}} + 0.4125IISP - 1.5652DISP$$
$$-\log NOEC(mM) = 0.3846(0) + 0.5312(1) + 0.0065(139.11)$$
$$+ 0.6155(0) + 0.4125(0.6) - 1.5652(0) = 1.683$$
(4.26*)

4.2 Organic Compounds

The development of industrialization is dangerous for natural flora and fauna. Organic contaminants have continuously been released into the surroundings. Organic compounds are classified into several subcategories based on their application and toxicological attributes, which include pesticides, pharmaceuticals, and cosmetics. The risk assessment of organic compounds is largely complex because different species react differently to different classes of organic compounds, surrounding structures, and targeted goals. Since large amounts of organic chemicals are synthesized and released into the environment, their toxicity can threaten aquatic life and the ecological system, especially causing human diseases such as gene damage and carcinogenicity [319]. Since toxicity assessment of organic compounds is of great importance for all chemical industries, some general QSAR/QSTR methods have been developed in recent years for organic compounds containing large data sets [320–327]. Several new QSAR/QSTR models based on different descriptors are reviewed here.

4.2.1 Chemical Toxicity to *Tetrahymena pyriformis* with Four Descriptor Models

Yu [328] developed a QSAR/QSTR model based on four descriptors for 1,163 chemical toxicants against *T. pyriformis* as follows:

$$-\log IC_{50}(mM) = 0.128 + 0.121ALOGP2 - 0.852GATS1p + 0.240MLIP + 0.002MW \quad (4.27*)$$

where *MW* is molecular weight, and the remaining three descriptors in eq. (**4.27***) are described as follows:

1. *ALOGP2*: The descriptor *ALOGP2* shows molecular properties that are the squared Ghose–Crippen–Viswanadhan octanol–water partition coefficient ($A \log K_{OW,GCV}$) and calculated by the following equation:

$$ALOGP2 = (A \log K_{OW,GCV})^2 = \left(\sum_i n_i a_i \right)^2 \qquad (4.28^*)$$

where n_i is the number of atoms i and a_i is the corresponding hydrophobicity constant.

2. *GATS1p*: The descriptor *GATS1p* is 2D autocorrelations that are Geary autocorrelations of lag 1 weighted by polarizability. It is calculated from the hydrogen-filled molecular graph by using the Geary coefficient which is a distance-type function varying from zero to infinite:

$$GATS1p = \frac{\frac{1}{2\Delta} \sum_{i=1}^{nsk} \sum_{j=1}^{nsk} \delta_{ij} (W_i - W_j)^2}{\frac{1}{nsk-1} \sum_{i=1}^{nsk} (W_i - \overline{W})^2} \qquad (4.29)$$

where W_i is the atomic property (polarizability); k is the lag value ($k = 1$), $\overline{W}$ is its average value of atomic polarizability; nsk is the number of atoms, δ_{ij} is the Kronecker delta ($\delta_{ij} = 1$ if $d_{ij} = k$, zero otherwise, d_{ij} being the topological distance between two considered atoms), Δ is the sum of the Kronecker deltas, that is, the number of atom pairs at distance equal to k. Thus, *GATS1p* describes how atomic polarizability is distributed along a topological molecular structure [320].

3. *MLIP*: The combinatorial descriptor *MLIP* reflects molecular lipophilicity, as well as molecular size and shape:

$$MLIP = n_{RNSC} + n_{DB} - n_{ROH}$$

where n_{RNSC}, n_{DB}, and n_{ROH} are the number of isothiocyanates, double bonds, and hydroxyl groups, respectively.

Although many factors influence the chemical toxicity to *T. pyriformis*, eq. (**4.27***) based on only four descriptors was successfully built up for $-\log IC_{50}(mM)$ of a large data set consisting of 1,163 compounds. Compared with previous QSAR/QSTR models of $-\log IC_{50}(mM)$ reported in the literature, eq. (**4.27***) shows satisfactory prediction performance, although eq. (**4.27***) has fewer descriptors.

4.2.2 Ecotoxicological QSAR/QSTR Modeling of Organic Compounds Against Fish

Khan et al. [321] developed QSAR/QSTR from a large data set of 1,121 organic compounds which can be used to predict the acute toxicity of organic ingredients in fish. They extracted fish mortality data (96 h LC_{50}, expressed as mg L^{-1}) for the whole set of 1,121 organic chemicals from merging six data sets available on the VEGA online platform (http://www.vegaqsar.eu/) with a great emphasis paid to homogeneous data collection to get reliable predictions. It should be mentioned that LD_{50} and LC_{50} are the parameters used to quantify the results of different tests so that they may be compared. LD_{50} is used for the dose that kills 50% of the test population but LC_{50} is used for the exposure concentration of a toxic substance lethal to half of the test animals. The mentioned data sets start from online repositories such as ECOTOX (https://cfpub.epa.gov/ecotox/), OPP (http://www.ipmcenters.org/ecotox/), QSAR Toolbox, AMBIT database (https://ambitlri.ideaconsult.net/), PPDB (https://sitem.herts.ac.uk/aeru/ppdb/en), Japanese Ministry of Environment (http://www.env.go.jp/en/chemi/sesaku/aquatic_Mar_2016.pdf; Japan Ministry of Environment, 2016), and integrating them with literature sources [329, 330]. They used SMILES to retrieve with automatic scripts and manually checked [331]. They obtained QSAR/QSTR models (local and global) using the PLS regression technique to obviate the effect of intercorrelation among the descriptors. They introduced modeling of the local data sets based on the specific functional group containing aldehydes, aliphatic amines, amides, anilines, esters, neutral organics, phenols, vinyl/allyl/propargyl (V/A/P) moiety-containing chemicals, and miscellaneous chemicals. They also provide modeling of the global data set using only DRAGON [332] and PaDEL descriptor [133] as well as predictivity enhancement using SiRMS descriptors (2D fragmental descriptors).

4.2.2.1 The QSAR/QSTR Modeling of the Local Data Sets
Khan et al. [321] derived nine QSAR models from individual classes, which are described in the following sections.

4.2.2.1.1 The QSAR/QSTR Model for Aldehydes

$$-\log LC_{50}(fish,\ mg\ L^{-1},\ 96\ h) = 2.765 + 0.119 MLOGP2 + 0.714 B08[C-C]$$
$$+ 1.010 B02[C-C] - 0.513 B05[C-C]$$
$$+ 0.523 Fr5(elm)/C\_C\_C\_H\_O$$
$$/1\_2s,\ 2\_3a,\ 3\_5s,\ 4\_5s/ + 0.151 F04[O-O] \qquad (4.30^*)$$

where $MLOGP2$ is squared Moriguchi octanol–water partition coefficient or $(\log K_{OW,M})^2$; $B08[C-C]$, $B02[C-C]$, and $B05[C-C]$ show the presence/absence of C–C at topological distances 8, 2, and 5, respectively; $Fr5(elm)/C\_C\_C\_H\_O/1\_2s,2\_3a,3\_5s,4\_5s/$ is differenti-

Table 4.5: Definitions of descriptors appearing in eq. (**4.30***) for aldehydes.

No.	Descriptors	Class of descriptor	Definition	Significance	Contribution in toxicity
1	B02[C–C]	2D atom pairs	Presence/absence of C–C at topological distance 2	Imparts hydrophobicity	Positive
2	B05[C–C]	2D atom pairs	Presence/absence of C–C at topological distance 5	Imparts hydrophobicity	Negative
3	B08[C–C]	2D atom pairs	Presence/absence of C–C at topological distance 8	Size and hydrophobicity	Positive
4	Fr5(elm)/ C_C_C_H_O/ 1_2s,2_3a,3_5s,4_5s/	SiRMS	Differentiated by elemental properties	Aromaticity	Positive
5	F04[O–O]	2D atom pairs	Frequency of O–O at topological distance 4	Imparts polarity	Positive
6	MLOGP2	Molecular properties	Squared Moriguchi octanol–water partition coefficient $(\log K_{OW,M})^2$	Imparts hydrophobicity	Positive

ated by elemental properties; $F04[O–O]$ is the frequency of O–O at topological distance 4. Table 4.5 gives definitions of descriptors appearing in eq. (**4.30***).

Toxicity of organic aldehydes has been reported in several works [333–335] in which $\log K_{OW}$ dependence of toxicity has been exhibited due to the presence of $MLOGP2$, $B08[C–C]$, and $B02[C–C]$. Two descriptors $B05[C–C]$ and $F04[O–O]$ contributed minimally toward fish toxicity. After lipophilicity, aromatic aldehyde was the most toxic fragment responsible for the organic chemical toxicity against fish.

4.2.2.1.2 The QSAR/QSTR Model for Aliphatic Amines

$$-\log LC_{50} \text{ (fish, mg L}^{-1}, 96 \text{ h)} = 1.442 - 0.635BLTD48 + 0.054ALogP2 + 0.34F03[C - S]$$
$$+ 0.205S\_A(chg)/A\_C\_C\_D/1\_2s, 1\_4s, 3\_4s/6$$
$$+ 0.674Fr5(d\_a)/A\_A\_A\_I\_I/1\_4s, 2\_5s, 3\_5d, 4\_5s/$$
$$- 0.079S\_A(rep)/B\_C\_C\_C/1\_3s, 1\_4s/4$$

$$(\textbf{4.31}^*)$$

where $BLTD48$ is Verhaar *Daphnia* baseline toxicity from $MLOGP$ (mmol L^{-1}) or $\log K_{OW,M}$; $ALogP2$ represents squared Ghose–Crippen octanol–water partition coefficient or $(A \log K_{OW,GCV})^2$; $F03[C–S]$ shows the frequency of C–S at topological distance 3; $S\_A(chg)/A\_C\_C\_D/1\_2s, 1\_4s, 3\_4s/6$ is differentiated by donor–acceptor groups;

Table 4.6: Definitions of descriptors appearing in eq. (**4.31***) for aliphatic amines.

No.	Descriptors	Class of descriptor	Definition	Significance	Contribution in toxicity
1	ALOGP2	Molecular properties	Squared Ghose–Crippen octanol–water partition coefficient ($A \log K_{ow,GCV}$)^2	Imparts hydrophobicity	Positive
2	BLTD48	Molecular properties	Verhaar *Daphnia* baseline toxicity from MLOGP (mmol L^{-1})	Index of lipophilicity	Negative
3	S_A(chg)/A_C_C_D/ 1_2s,1_4s,3_4s/6	SiRMS	Differentiated by charge	Given in [321]	Positive
4	Fr5(d_a)/A_A_A_I_I/ 1_4s,2_5s,3_5d,4_5s/	SiRMS	Differentiated by donor–acceptor groups	Given in [321]	Positive
5	F03[C-S]	2D atom pairs	Frequency of C–S at topological distance 3	Size and polarity	Positive
6	S_A(rep)/B_C_C_C/ 1_3s,1_4s/4	SiRMS	Differentiated by repulsion properties		Negative

S_A(rep)/B_C_C_C/1_3s, 1_4s/4 is differentiated by repulsion properties. Table 4.6 provides definitions of descriptors appearing in eq. (**4.31***).

Since the baseline or minimal toxic concentration against *Daphnia* can exert maximum contribution in regulating fish toxicity, it was found to be the most significant descriptor present in eq. (**4.31***). Due to the negative correlation of BLTD48, there is a simultaneous linear increase in the potency of aliphatic amines against fish as well as *Daphnia*. Higher concentrations of aliphatic amines can cause mortality in crustaceans, which provide safer molecule against fish. The descriptor AlogP2 also enhanced fish toxicity significantly. Three descriptors F03[C-S], S_A(chg)/A_C_C_D/1_2 s,1_4 s,3_4 s/6, and Fr5(d_a)/A_A_A_I_I/1_4 s,2_5 s,3_5d,4_5 s/ contributed positively, while S_A(rep)/ B_C_C_C/1_3 s,1_4 s/4 has a negative impact on fish mortality insignificantly.

4.2.2.1.3 The QSAR/QSTR Model for Amides

$$-\log LC_{50} \text{ (fish, mg L}^{-1}, 96 h) = 0.362 + 0.554X1sol + 0.763B07[C-N] + 0.992F02[C-S]$$

$$-0.423F03[C-S] - 0.094H\text{-}047$$

(**4.32***)

where *X1sol* is the solvation connectivity index of order 1; *B07[C–N]* represents the presence/absence of C–N at topological distance 7; *F02[C–S]* and *F03[C–S]* show the frequency of C–S at topological distances 2 and 3, respectively; *H-047* is H attached to C1(sp^3)/C0(sp^2). Table 4.7 shows definitions of descriptors appearing in eq. (**4.32***).

Table 4.7: Definitions of descriptors appearing in eq. (**4.32***) for amides.

No.	Descriptors	Class of descriptor	Definition	Significance	Contribution in toxicity
1	F03[C-S]	2D atom pairs	Frequency of C–S at topological distance 3	Size and polarity	Negative
2	B07[C-N]	2D atom pairs	Presence/absence of C–N at topological distance 7	Size and polarity	Positive
3	F02[C-S]	2D atom pairs	Frequency of C–S at topological distance 2	Size and polarity	Positive
4	H-047	Atom-centered fragments	H attached to C1(sp^3)/CO(sp^2)	Size and polarity	Negative
5	X1sol	Connectivity indices	Solvation connectivity index of order 1	Imparts solvation property in the molecule	Positive

Due to the presence of X1sol variable in eq. (**4.32***), toxicity increases with an increment in the solvation property of molecules. The presence of nitrogen in the form of *B07[C–N]* enhances the toxicity of fish despite increasing polarity. The existence of sulfur depends on its position because of the presence of sulfur at the topological distance 2 from carbon-enhanced fish mortality, whereas with a longer chain toxicity decreases. The descriptor *H-047* contributed negatively but insignificantly.

4.2.2.1.4 The QSAR/QSTR Model for Anilines

$$-\log LC_{50} \, (fish, \, mg\,L^{-1}, \, 96\,h) = 2.888 + 0.208 ALogP2 + 0.068 MLOGP2 - 0.018 T(N \ldots F)$$
$$+ \, 0.096 S\_A(att)/E\_E\_E - F/1\_4s, \, 2\_3a/3 - 0.275H - 051$$
$$+ \, 0.188 Fr5(lip)/B\_B\_B\_C\_C/1\_2s, \, 2\_4a, \, 3\_5a, \, 4\_5a/$$
$$+ \, 0.136 F02[C - N] + 0.209 F07[O - O]$$

$$(4.33^*)$$

where *ALogP2* represents the squared Ghose–Crippen octanol–water partition coefficient or $(A \log K_{OW,GCV})^2$; *MLOGP2* is squared Moriguchi octanol–water partition coefficient or $(\log K_{OW,M})^2$; $T(N \ldots F)$ is the sum of topological distances between $N \ldots F$; *S_A(att)/E_E_E-F/1_4s, 2_3a/3-0.275H-051* is differentiated by attraction properties; *Fr5(lip)/B_B_B_C_C/1_2s, 2_4a, 3_5a, 4_5a/* is differentiated by lipophilic properties; *F02[C–N]* is the frequency of C–N at topological distance 2; *F07[O–O]* is the frequency of O–O at topological distance 7. Table 4.8 gives definitions of descriptors appearing in eq. (**4.33***).

Table 4.8: Definitions of descriptors appearing in eq. (**4.33***) for anilines.

No.	Descriptors	Class of descriptor	Definition	Significance	Contribution in toxicity
1	ALOGP2	Molecular properties	Squared Ghose–Crippen octanol–water partition coefficient $(A \log K_{OW,GCV})^2$	Imparts hydrophobicity	Positive
2	MLOGP2	Molecular properties	Squared Moriguchi octanol–water partition coefficient $(\log K_{OW,M})^2$	Imparts hydrophobicity	Positive
3	S_A(att)/E_E_E_F/ 1_4s,2_3a/3	SiRMS	Differentiated by attraction properties	Given in [321]	Positive
4	F02[C–N]	2D Atom pairs	Frequency of C–N at topological distance 2	Size and polarity	Positive
5	F07[O–O]	2D Atom pairs	Frequency of O–O at topological distance 7	Imparts polarity	Positive
6	H-051	Atom-centered fragments	H attached to alpha-C	Size and polarity	Negative
7	Fr5(lip)/B_B_B_C_C/ 1_2s,2_4a,3_5a,4_5a/	SiRMS	Differentiated by lipophilic properties	Given in [321]	Positive
8	T(N..F)	2D atom pairs	The sum of topological distances between N. . .F	Imparts polarity	Negative

Due to the presence of two descriptors *ALogP2* and *MLOGP2* in eq. (**4.33***), aniline derivatives show lipophilicity. Moreover, eq. (**4.33***) also focuses on the influence of aromatic rings (mainly aniline and chlorobenzene) in defining organic substance toxicity in fish. Due to the positive coefficients of *S_A(att)/E_E_E_F/1_4 s,2_3a/3* and *Fr5(lip)/B_B_B_C_C/1_2 s,2_4a,3_5a,4_5a/*, eq. (**4.33***) mainly focuses on aromaticity dependence of the toxicity. Two descriptors *F02[C–N]* and *F07[O–O]* contributed very less in defining the organic chemical.

4.2.2.1.5 The QSAR/QSTR Model for Esters

$$-\log LC_{50} \ (fish, \ mg \ L^{-1}, \ 96 \ h) = -1.862 + 0.313ALOGP + 0.044D/Dtr03 + 0.768B06[C-O]$$
$$+ 7.521Mp - 0.552Fr5(lip)/A\_C\_C\_C\_D/1\_2s, 1\_3s, 4\_5s/$$
$$- 0.406F09[C-S] + 0.304F03[O-P]$$

$$(4.34^*)$$

where *ALOGP* represents the Ghose–Crippen octanol–water partition coefficient or ($A \log K_{OW,GCV}$); *D/Dtr03* is the distance/detour ring index of order 3; *B06[C–O]* is the presence/absence of C–O at the topological distance 6; Mp is mean atomic polarizability (scaled on carbon atom); $Fr5(lip)/A\_C\_C\_C\_D/1\_2s, 1\_3s, 4\_5s/$ is differentiated by lipophilic properties; $F09[C–S]$ is the frequency of C–S at topological distance 9; $F03[O–P]$ gives the frequency of O–P at topological distance 3. Table 4.9 provides definitions of descriptors appearing in eq. (**4.34***).

Table 4.9: Definitions of descriptors appearing in eq. (**4.34***) for esters.

No.	Descriptors	Class of descriptor	Definition	Significance	Contribution in toxicity
1	ALOGP	Molecular properties	Ghose–Crippen octanol–water partition coefficient (A log K_{OW}, GCV)	Imparts hydrophobicity	Positive
2	B06[C–O]	2D atom pairs	Presence/absence of C–O at topological distance 6	Size and polarity	Positive
3	D/Dtr03	Ring descriptors	Distance/detour ring index of order 3	Imparts size	Positive
4	F03[O–P]	2D atom pairs	Frequency of O–P at topological distance 3	Imparts polarity	Positive
5	F09[C–S]	2D atom pairs	Frequency of C–S at topological distance 9	Size and polarity	Negative
6	Fr5(lip)/A_C_C_C_D/ 1_2s,1_3s,1_5s,4_5s/	SiRMS	Differentiated by lipophilic properties	Given in [321]	Negative
7	Mp	Constitutional indices	Mean atomic polarizability (scaled on carbon atom)	Imparts polarizability	Positive

Equation (**4.34***) suggests a direct correlation between the lipophilicity of organic molecules the fish mortality. Four descriptors ALOGP, D/Dtr03, B06[C–O], and Mp have positive coefficients, which contribute most significantly to controlling fish mortality due to organic chemicals. Two variables Fr5(lip)/A_C_C_C_D/1_2s, 1_3s, 1_5s, 4_5s/ and F09[C–S] contribute negatively in controlling fish mortality. Meanwhile, F03[O–P] exerted a positive contribution to fish toxicity, insignificantly.

4.2.2.1.6 The QSAR/QSTR Model for Neutral Organics

$$-\log LC_{50}\ (fish,\ mg\ L^{-1},\ 96\ h) = 0.638 + 0.823ALOGP + 0.043O\% + 0.269Ui$$
$$+ 0.007TPSA(Tot) + 1.509ETA\_Psi\_1 - 0.131F05[C - O]$$
$$- 0.634B02[C - C] + 0.852X3Av$$

$$(4.35^*)$$

where *ALOGP* represents Ghose–Crippen octanol–water partition coefficient or (*A* log $K_{OW,GCV}$); O% is the percentage of O atoms; *Ui* is unsaturation index; *TPSA(Tot)* is topological polar surface area using N, O, S, P polar contributions; *ETA_Psi_1* is a measure of hydrogen bonding propensity of the molecules and/or polar surface area; *F05[C–O]* is the frequency of C–O at topological distance 5; *B02[C–C]* gives the presence/absence of C–C at topological distance 2; *X3Av* shows the average valence connectivity index of order 3. Table 4.10 offers definitions of descriptors appearing in eq. (**4.35***).

Table 4.10: Definitions of descriptors appearing in eq. (**4.35***) for neutral organics.

No.	Descriptors	Class of descriptor	Definition	Significance	Contribution in toxicity
1	*ALOGP*	Molecular properties	Ghose–Crippen octanol–water partition coefficient (*A* log $K_{OW,GCV}$)	Imparts hydrophobicity	Positive
2	*B02[C–C]*	2D atom pairs	Presence/absence of C–C at topological distance 2	Imparts hydrophobicity	Negative
3	*ETA_Psi_1*	ETA indices	A measure of hydrogen bonding propensity of the molecules and/ or polar surface area	The measure of hydrogen bond donor or acceptor	Positive
4	*F05[C–O]*	2D atom pairs	Frequency of C–O at topological distance 5	Size and polarity	Negative
5	*O%*	Constitutional indices	Percentage of O atoms	Polar bulk	Positive
6	*TPSA(Tot)*	Molecular properties	Topological polar surface area using N, O, S, P polar contributions	Imparts polarity	Positive
7	*Ui*	Molecular properties	Unsaturation index	Imparts hydrophobicity	Positive
8	*X3Av*	Connectivity indices	Average valence connectivity index of order 3	Hydrophobic content, Bulk, Size	Positive

Since eq. (**4.35***) was obtained from 374 organic chemicals, it offers a larger chemical and biological domain for its application in the ecotoxicological modeling of the fish toxicity endpoint. Two descriptors *ALOGP* and *X3Av* with positive coefficients enhance fish mortality because they have lipophilic moieties. *Ui* also suggests that the toxicity of organic chemicals increases with an increment of their unsaturation. Due to the dependence of toxicity from descriptors like *O%*, *TPSA(Tot)*, and *ETA_Psi_1*, the idea of polar narcosis can be confirmed to some extent [336]. Two descriptors *F05[C-O]* and *B02[C-C]* with negative coefficients hurt fish mortality.

4.2.2.1.7 The QSAR/QSTR Model for Phenols

$$-\log LC_{50} \left(fish,\ mg\ L^{-1},\ 96\ h\right) = 1.0162 - 0.927BLTA96 + 0.940n_{Crs}$$
$$- 0.130Fr5(lip)/B\_C\_C\_C\_C/1\_3a,\ 2\_3s,\ 3\_5a,\ 4\_5a/$$
$$+ 0.411n_{ArNO2} + 0.995N\text{-}069 + 0.647n_{ArCHO}$$

$$(4.36^*)$$

where *BLTA96* represents Verhaar algae baseline toxicity from *MLOGP* (mmol L^{-1}) or $\log K_{OW, M}$; n_{Crs} is the number of ring secondary C(sp^3); $Fr5(lip)/B\_C\_C\_C\_C/1\_3a,\ 2\_3s,$ $3\_5a,\ 4\_5a/$ is differentiated by lipophilic properties; n_{ArNO2} is the number of nitro groups (aromatic); *N-069* is the presence of Ar-NH$_2$/X-NH$_2$ where X can be O, N, S, P, Se, and halogens; n_{ArCHO} is the number of aldehydes (aromatic). Table 4.11 gives definitions of descriptors appearing in eq. (**4.36***).

Table 4.11: Definitions of descriptors appearing in eq. (**4.36***) for phenols.

No.	Descriptors	Class of descriptor	Definition	Significance	Contribution in toxicity
1	BLTA96	Molecular properties	Verhaar Algae base-line toxicity from MLOGP (mmol L^{-1}) or $\log K_{OW, M}$	Index of lipophilicity	Negative
2	Fr5(lip)/B_C_C_C_C/ 1_3a,2_3s,3_5a,4_5a/	SiRMS	Differentiated by lipophilic properties	Given in [321]	Negative
3	N-069	Atom-centered fragments	Presence of Ar-NH$_2$/X-NH$_2$, where X can be O, N, S, P, Se, and halogens	Aromaticity	Positive
4	n_{ArCHO}	Functional group counts	Number of aldehydes (aromatic)	Imparts polarity	Positive
5	n_{ArNO2}	Functional group counts	Number of nitro groups (aromatic)	Imparts polarity	Positive
6	n_{Crs}	Functional group counts	Number of ring secondary C (sp^3)	Imparts hydrophobicity	Positive

Among six descriptors given in eq. (**4.36***), *BLTA96* and $Fr5(lip)/B\_C\_C\_C\_C/1\_3a,2\_3$ $s,3\_5a,4\_5a/$ contributed negatively while the remaining four descriptors namely, n_{Crs}, n_{ArNO2}, *N-069*, and n_{ArCHO} contributed positively to the fish toxicity of organic chemicals. Since *BLTA96* has a negative coefficient, it suggests a linear decrease in toxicity of organic chemicals against fish with a rise in the baseline toxicity concentration against algae. Highly lipophilic molecules can have higher fish toxicity because baseline toxicity concentration is inversely dependent on the lipophilicity of the molecules.

4.2.2.1.8 The QSAR/QSTR Model for Vinyl/Allyl/Propargyl (V/A/P) Moiety Containing Chemicals

$$-\log LC_{50}\,(fish,\, mg\,L^{-1},\; 96\,h) = 3.415 + 0.169X0sol + 0.048S\_A(lip)/B\_B\_C\_C/2\_4s,\, 3\_4s/4$$
$$- 2.524O - 057 - 0.226Fr5(en)/B\_B\_C\_C\_D/1\_4s,\, 2\_4s,\, 3\_4s,$$
$$3\_5s/ - 1.008n_{OHt} + 1.668B03[N - Cl] - 1.726B04[N - O]$$

$$(4.37^*)$$

where *X0sol* represents solvation connectivity index of order 0; *S_A(lip)/B_B_C_C/2_4s, 3_4s/4* is differentiated by lipophilic properties; *Fr5(en)/B_B_C_C_D/1_4s, 2_4s, 3_4s, 3_5s/* is differentiated by electronegativity; n_{OHt} is the number of tertiary alcohols; *B03[N–Cl]* is the presence/absence of N–Cl at topological distance 3; *B04[N–O]* is the presence/absence of N–O at topological distance 4. Table 4.12 provides definitions of descriptors appearing in eq. (**4.37***).

Table 4.12: Definitions of descriptors appearing in eq. (**4.37***) for V/A/P.

No.	Descriptors	Class of descriptor	Definition	Significance	Contribution in toxicity
1	B03[N-Cl]	2D atom pairs	Presence/absence of N–Cl at topological distance 3	Imparts polarity	Positive
2	B04[N-O]	2D atom pairs	Presence/absence of N–O at topological distance 4	Imparts polarity	Negative
3	Fr5(en)/B_B_C_C_D/ 1_4s,2_4s,3_4s,3_5s/	SiRMS	Differentiated by electronegativity	Given in [321]	Negative
4	S_A(lip)/B_B_C_C/ 2_4s,3_4s/4	SiRMS	Differentiated by lipophilic properties	Given in [321]	Positive
5	n_{OHt}	Functional group counts	Number of tertiary alcohols	Imparts polarity	Negative
6	O-057	Atom-centered fragments	Phenol/enol/carboxyl OH	Imparts polarity	Negative
7	X0sol	Connectivity indices	Solvation connectivity index of order 0	Imparts solvation property in the molecule	Positive

Like eq. (**4.32***), in the VAP data set, the *X0sol* variable exerted a positive influence on fish mortality. Due to the positive coefficients of *S_A(lip)/B_B_C_C/2_4 s,3_4 s/4*, and *B03[N–Cl]*, they have a lethal effect on fish. Since four variables *O-057, Fr5(en)/*

$B\_B\_C\_C\_D/1\_4s$, 2_4s, 3_4s, $3\_5s/$, n_{OHt} and $B04[N–O]$ have positive coefficients, they have contributed negatively in fish mortality of V/A/P organic chemicals.

4.2.2.1.9 The QSAR/QSTR Model for Miscellaneous Chemicals

$$-\log LC_{50}\,(fish,\ mg\ L^{-1},\ 96\ h) = -1.606 - 0.530BLTF96 + 0.251X5sol + 10.594X1A$$
$$+ 0.027D/Dtr03 - 0.197F02[N-N] + 0.551O - 060$$
$$- 0.532B01[C-O] - 0.486S\_A(type)/C.1\_C.1\_C.3\_H/$$
$$2\_3s,\ 3\_4s/4 + 0.786NssS - 0.021H\%$$

$$(4.38^*)$$

where $BLTF96$ represents Verhaar Fish base-line toxicity from $MLOGP$ (mmol/l); $X5sol$ is solvation connectivity index of order 5; $X1A$ is average connectivity index of order 1; $D/Dtr03$ is distance/detour ring index of order 3; O-060 is Al-O-Ar /Ar-O-Ar/R..O..R/R-O-C = X; $B01[C-O]$ is presence/absence of C-O at topological distance 1; $S\_A(type)/C.\ 1\_C.$ $1\_C.3\_H/2\_3s$, $3\_4s/4$ is differentiated by elemental properties; $H\%$ is percentage of H atoms. Table 4.13 offers definitions of descriptors appearing in eq. (4.38*).

Table 4.13: Definitions of descriptors appearing in eq. (4.38*) for miscellaneous chemicals.

No.	Descriptors	Class of descriptor	Definition	Significance	Contribution in toxicity
1	D/Dtr03	Ring descriptors	Distance/detour ring index of order 3	Imparts size	Positive
2	B01[C-O]	2D atom pairs	Presence/absence of C–O at topological distance 1	Imparts polarity	Negative
3	BLTF96	Molecular properties	Verhaar fish baseline toxicity from MLOGP (mmol L^{-1})	Index of lipophilicity	Negative
4	F02[N-N]	2D atom pairs	Frequency of N–N at topological distance 2	Imparts polarity	Negative
5	H%	Constitutional indices	Percentage of H atoms	Imparts polarity	Negative
6	NssS	atom-type E-state indices	Number of atoms of type ssS	Imparts polarity	Positive
7	O-060	Atom-centered fragments	Al-O-Ar/Ar-O-Ar/R. . .O. . .R/R-O-C=X	Presence of ether moiety	Positive

Table 4.13 (continued)

No.	Descriptors	Class of descriptor	Definition	Significance	Contribution in toxicity
8	S_A(type)/ C.1_C.1_C.3_H/ 2_3s,3_4s/4	SiRMS	Differentiated by elemental properties	Given in [321]	Negative
9	X1A	Connectivity indices	Average connectivity index of order 1	Hydrophobic content, bulk, size	Positive
10	X5sol	Connectivity indices	Solvation connectivity index of order 5	Imparts solvation property in molecule and size	Positive

The quality of eq. (**4.38***) for miscellaneous chemicals was inferior to the rest of the other models in terms of validation metrics, which may be due to the very different chemical classes present in the data set. Various groups were used for deriving this model in the miscellaneous data sets, which include acrylates, acrylamides, benzodioxoles, benzyl halides, benzyl alcohols, carbonyl ureas, haloacids, epoxides, haloketones, hydrazines, imidazoles, hydroquinones, imides, methacrylates, malonitriles, nitriles, phthalonitriles, pyrazoles/pyrroles, and thiols. Equation (**4.38***) consists of 10 theoretical descriptors, 5 of which contributed positively, and 5 variables negatively toward fish toxicity. The variable BLTF96 with a negative coefficient has the most significant descriptor present in the model suggesting a linear (inverse) correlation with the toxicity of miscellaneous chemicals against fish. *X5sol, X1A, D/Dtr03, O-060*, and *NssS* descriptors enhance the fish mortality of the miscellaneous chemical class data set. Meanwhile, *F02[N-N], B01[C-O], S_A(type)/C.1_C.1_C.3_H/2_3 s,3_4 s/4*, and *H%* are negatively correlated descriptors.

4.2.2.2 The QSAR/QSTR Modeling of the Global Data Set

4.2.2.2.1 Using Only Dragon and PaDEL Descriptor Software

$$-\log LC_{50} \text{ (fish, } mg\, L^{-1}, \; 96\, h) = 5.364 + 0.431 ALOGP - 0.002 MW + 0.233 X1sol - 0.028 H\%$$
$$+ 1.0810 F01[C-X] - 2.908 ETA\_BetaP\_S + 0.777 B10[C-N]$$
$$- 0.151 n_{R06} - 0.271 n_{ROH} + 0.354 C - 015$$

$$(\mathbf{4.39^*})$$

where *ALOGP* represents lipophilicity; *MW* is the molecular weight; *X1sol* is the solvation connectivity index of order 1; *H%* is the percentage of H atoms (constitutional index); *F01[C–X]* is the frequency of carbon and metal elements at topological distance 1; *ETA_BetaP_S* is a measure of electronegative atom count of the molecule relative to molecular size; *B10[C–N]* is the presence of carbon and nitrogen at the topological distance 10 (2D atom pairs); n_{R06} is number of six-membered rings; n_{ROH} is number of hydroxyl groups (functional group count); *C-015* is $=CH_2$ (atom-centered fragment).

Equation (**8.39***) for the whole data set employing Dragon and PaDEL descriptor-derived descriptors exhibited moderate robustness for a very large data set of 1,121 molecules. Among different descriptors in eq. (**8.39***), three descriptors *ALOGP*, *MW*, and *X1sol* contributed maximum toward fish toxicity. Meanwhile, seven descriptors *H%*, *F01[C-X]*, *ETA_BetaP_s*, *B10[C–N]*, n_{R06}, n_{ROH}, and *C-015* are the less significant descriptors.

4.2.2.2.2 Using SiRMS Descriptors (2D Fragmental Descriptors)

$$-\log LC_{50}\ (fish,\ mg\ L^{-1},\ 96\ h) = 1.347 + 0.420ALOGP + 0.019XMOD + 1.801Mv$$
$$- 0.327S\_A(type)/C.3\_C.3\_H\_O.3/1\_2s,\ 2\_4s,\ 3\_4s/6$$
$$+ 0.738B10[C - N] + 0.162Fr3(rf)/A\_B\_B/1\_2s,\ 2\_3d/$$
$$- 0.384Fr3(type)/C.3\_C.3\_O.3/1\_3s,\ 2\_3s/$$
$$+ 0.873F01[C - X] - 0.352Fr3(rep)/B\_D\_E/1\_3s,\ 2\_3d/$$
$$- 0.254Fr3(att)/D\_D\_E/1\_3s,\ 2\_3s/$$

$$(4.40^*)$$

where *Mv* is the mean atomic van der Waals volume (scaled on carbon atom); *XMOD* gives the modified Randic index; employing fragmental descriptors consisting of five SiRMS descriptors namely *Fr3(att)/D_D_E/1_3s, 2_3s/*, *Fr3(rf)/A_B_B/1_2s, 2_3d/*, *Fr3(type)/ C. 3_C. 3_O. 3/1_3s, 2_3s/*, *S_A(type)/C. 3_C. 3_H_O. 3/1_2s, 2_4s, 3_4s/6*, and *Fr3(rep)/B_D_E/ 1_3s, 2_3d/* are differentiated by different atomic features like attraction, refraction, atom type, and repulsion, respectively. Due to employing SiRMS features, eq. (**4.40***) has slightly enhanced the predictive quality with the extent of complexity in model development.

4.2.2.2.3 Some Features of Descriptors in Equations (4.39*) and (4.40*)

Equations (**4.39***) and (**4.40***) consist of 17 theoretical descriptors including five 2D simplex molecular fragments. The modeled descriptors can simply be grouped into two categories as eight descriptors contributed positively toward the fish toxicity while the rest of nine descriptors contributed negatively.

Several descriptors are related to log K_{OW} dependence on aquatic toxicity of environmental contaminants, which enhance the hydrophobic features of a molecule directly or indirectly. They are *ALOGP*, *X1sol*, *Mv*, *C-015*, and *XMOD*. *ALOGP* is the most significant descriptor present in eqs. (**4.39***) and (**4.40***). The descriptor *ALOGP* describes the degree of hydrophobicity in a molecule, which can be calculated mainly for C, H, N, O, S, Se, P, B, Si, and halogen atoms. A hydrophilic moiety present in molecules is the second major feature contributing to fish toxicity. It can be represented by descriptors such as $H\%$, n_{ROH}, ETA_BetaP_s, and $B10[C–N]$. Higher electronegative atom count (ETA indices such as ETA_BetaP_s) has a mixed response to the fish toxicity of organic compounds. The presence of oxygen mainly in the form of hydroxyl and ether groups reduces fish toxicity while toxicity due to nitrogen had a mixed effect based on its position.

The SiRMS descriptors are a series of tetratomic fragments identified in a molecule, which can identify the probable feature responsible for fish toxicity. Equations **(4.39*)** and **(4.40*)** with SiRMS descriptor contained five such repetitive fragments. The descriptor $S\_A(type)/C.3\_C.3\_H\_O.3/1\_2\ s,2\_4\ s,3\_4\ s/6$ is an atom-type SiRMS-based descriptor, corresponds to the following fragment: $C(sp^3)$-$C(sp^3)$-H-O(sp^3). It measures the aliphatic hydroxyl fragment present in a molecule. Since it has a negative contribution, the presence of a higher number of such fragments reduces toxicity toward fish species. The descriptor $Fr3(type)/C.3\_C.3\_O.3/1\_3\ s,2\_3\ s/$ has a negative contribution to the fish mortality to the presence of atom-type three-atomic fragment, which is an ether. The fragment may be either noncyclic or cyclic. The descriptor $Fr3(rf)/A\_B\_B/1\_2\ s,2\_3d/$ has a positive effect on the fish toxicity, contributed by the refraction of four atomic fragments such as the vinyl group. The descriptor $Fr3(att)/D\_D\_E/1\_3\ s,2\_3\ s/$ gives the count of heterocyclic aromatic rings or urea/thiourea/N-urea moiety that has a negative contribution to the fish toxicity. The descriptor $Fr3(rep)/B\_D\_E/1\_3\ s,2\_3d/$ represents a four-atomic fragment such as $-P(=O)$-$C(CH_3)$- or $-S(=O)$-$C(CH_3)$-, which has a negative contribution to the fish toxicity.

4.2.2.2.4 Overall Interpretation and Application of Equations (4.30*)–(4.40*)

The greatest of the local models support log K_{OW} dependence on the toxicity of organic compounds. Lipophilic contributions from halogens (mainly chlorine) and ring aromatic features are other crucial features responsible for fish mortality. Toxic fragments are anilines, aldehydes, chlorobenzene, sulfur, and nitrogen. Higher baseline toxicity concentration, long chains, and polar functionalities like oxygen can reduce the toxicity potential of organic compounds.

For global models including eqs. **(4.39*)** and **(4.40*)**, $ALOGP$, cyanide, and vinyl group are the most significant features of organic chemicals, which enhance organic chemical toxicity. Meanwhile, aliphatic ethers, hydroxyl, and polar groups like oxygen contributed negatively to enhancing the toxicity of organic compounds. The presence of nitrogen has a mixed effect, which is based on its position in the molecule.

For QSAR/QSTR models of aldehydes, aliphatic amines, anilines, esters, neutral organics, and both the global models including eqs. **(4.39*)** and **(4.40*)**, log K_{OW} (log P) is a more influential variable, which reflects the potential of chemicals to participate in hydrophobic interactions with lipoidal tissues in various fish species and to permeate through the lipoidal membrane [337]. Aldehyde moieties are highly toxic in smaller concentrations than their nonreactive counterparts of equal hydrophobicity because of their covalent binding affinity toward nucleophilic groups in proteins and nucleic acids, which form Schiff bases with amino groups in biological macromolecules [333]. Phenol derivatives can accumulate in different organs of fish, resulting in different physiological problems or death depending upon their concentration [338]. They may enter into the blood circulation of fish from water through gills or mucous epithelium of the mouth [338]. Thus, the acute aquatic toxicity of organic chemicals against fish can follow mainly nonpolar pathways. Moreover, nonpolar narcotics are neutral nonreactive organic chemicals such as aliphatic alcohols, ethers, and ketones whose toxic effect is mainly governed by log K_{OW} [339]. Thus, eqs. (**4.30***) and (**4.40***) confirm the established fish toxicity mechanisms as observed in the previous works [337–339].

4.3 Summary

This chapter demonstrated several methods for assessments of toxicities of aromatic and organic compounds in Sections 4.1 and 4.2. Five eqs. (**4.1***)–(**4.5***) provided five models containing a defined endpoint ($-\log IC_{50}(mM)$) against *T. pyriformis* employing a reduced pool of descriptors that have R^2 values in the range of 0.737–0.740. They have descriptors containing negative or positive contributions toward *T. pyriformis*, which are given in Table 4.1 as well as in Sections 4.1.1.1 and 4.1.1.2. Equation (**4.6***) can predict the acute toxicity endpoint ($-\log LC_{50}$) of aromatic chemicals in tadpoles of the Japanese brown frog (*R. japonica*) using correlation weights. Equations (**4.7***)–(**4.11***) introduced five QSAR/QSTR models to predict acute toxicity ($-\log LC_{50}$) of different aromatic compounds against the tadpoles of *R. japonica*. Section 4.1.4 introduced eqs. (**4.13***), (**4.16***), and (**4.17***) that were used to identify and predict the acute toxicity ($-\log IC_{50}$) of substituted aromatic compounds, which include compounds with nitro groups, halogen substituents, and both nitro and halogen substituents, respectively, to the aquatic ciliate *T. pyriformis*. Equations (**4.19***), (**4.20***), and (**4.21***) were used to correlate $-\log EC_{50}$, $-\log TU_0$, and $-\log TU_{1/2}$ with structural features of studied S-BRCs over molecular descriptors for predicting the toxicity toward *V. fischeri* of aromatics and their reaction mixtures during the treatment by AOPs. Equation (**4.22***) is a simple approach for reliable prediction of the toxicity of organic aromatic compounds based on endpoint $-\log IC_{50}(mM)$ toward *T. pyriformis*. It requires only structural parameters of a desired aromatic compound without using complex descriptors. Equations (**4.23***), (**4.24***), (**4.25***), and (**4.26***) used the same five simple structural descriptors for finding endpoints $-\log IC_{50}$, $-\log IC_{20}$, $-\log LOEC$, and $-\log NOEC$ toward *C. vulgaris* of aromatics, respectively. Equation (4.27*) is a QSAR/QSTR model based on four descriptors for the assessment of toxicity of organic compounds against *T. pyriformis*. Equations (**4.30***), (**4.31***), (**4.32***), (**4.33***), (**4.34***), (**4.35***), (**4.36***), (**4.37***), and (**4.38***) introduced different QSAR/QSPR models for the prediction of $-\log LC_{50}$ *(fish, mg L^{-1}, 96 h)* of aldehydes, aliphatic amines, amides, anilines, esters, neutral organics, phenols, V/A/P moiety containing chemicals, and miscellaneous chemicals, respectively. Equations (**4.39***) and

(**4.40***) are two global QSAR/QSTR modeling of organic compounds. Equation (**4.39***) is based on Dragon and PaDEL descriptor software. Meanwhile, (**4.40***) requires *SiRMS* descriptors (2D-fragmental descriptors).

Problems

1. 4-Phenoxyphenol has the following molecular structure:

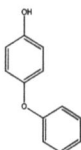

(a) Calculate $\sum CW(S_k)$ from SMILES. (a) Use eq. (**4.6***) to calculate $-\log LC_{50}$. (b) If the experimental value of $-\log LC_{50}$ is 4.0300, calculate the deviation of eq. (**4.6***).

2. 4-Bromo-2-fluoro-6-nitrophenol has the following molecular structure:

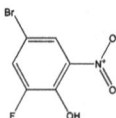

(a) Use eq. (**4.22***) to calculate $-\log IC_{50}(mM)$. (b) If the experimental value of $-\log IC_{50}(mM)$ is 1.62, calculate the deviation of eq. (**4.22***). (c) If the predicted result of the method of Su et al. [308] for this compound is 1.35, which method gives closer output as compared to the measured value?

3. Using eqs. (**4.23***) to (**4.26***), calculate the values of $-\log IC_{50}$, $-\log IC_{20}$, $-\log LOEC$, and $-\log NOEC$ toward *C. vulgaris* for 4-chloro-3,5-dimethylphenol with the following molecular structure:

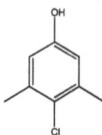

Chapter 5
Toxicity of Ionic Liquids

Ionic liquids (ILs) are salts comprising cations and anions with a melting point below 100 °C that have wide applications [340, 341]. Since ILs are nonvolatile, this characteristic is one of the most important properties for making them potentially "green" alternatives to volatile organic compounds. ILs can be released into aquatic ecosystems because many ILs are soluble in water and contribute to water pollution. Toxicities of some ILs have been assessed in the literature but a great numbers of ILs have been synthesized each year. Thus, the available information on toxicity is still scarce. Determination of the toxicity of ILs is important by an inhibition assay, which can assess the toxicological danger of ILs. It is valuable to educate on the impact of ILs on all classes because both water- and non-water-soluble ILs can be toxic, that is, hydrophobic/hydrophilic ones [342].

Many studies have described the toxicity of diverse IL subfamilies against diverse organisms' toxicity such as esterase enzyme, *Vibrio fischeri*, algae, and *Daphnia magna*. Abramenko et al. [343] reviewed some advances toward the development of QSAR/QSTR models for the toxicity assessment of ILs. For example, several works have been done to estimate the toxicity of ILs based on *V. fischeri* [344–347]. Alvarez-Guerra and Irabien [348] as well as Das and Roy [347] developed two different QSAR/QSTR models to estimate the class of toxicity to *V. fischeri* from a relatively large database. Yan et al. [344] derived a QSAR/QSTR technique using toxicity data of *V. fischeri*, which are composed of 74 cations and 22 anions. Tables 5.1 and 5.2 show the chemical formula, abbreviation, and structure of some common cations and anions, respectively.

Table 5.1: Chemical formula, abbreviation, and structure of some common cations.

Chemical formula	Abbreviation	Structure
$C_2H_8NO^+$	[2-HEA]	
$C_3H_5N_2^+$	[IM]	
$C_3H_7N_6^+$	[Melanime]	
$C_4H_6NO^+$	[MOxa]	

https://doi.org/10.1515/9783111189673-005

Table 5.1 (continued)

Chemical formula	Abbreviation	Structure
$C_4H_7N_2^+$	[MIM]	
$C_4H_{12}NO_2^+$	[2-HDEA]	
$C_5H_9N_2^+$	[DMIM]	
$C_5H_9N_2^+$	[EIM]	
$C_5H_{14}NO^+$	[Choline]	
$C_5H_{14}N_3^+$	[TMG]	
$C_6H_8N_3^+$	[MIMCN] or [CNC$_1$mim]	
$C_6H_{11}N_2^+$	[EMIM] or [C$_2$mim]	
$C_6H_{11}N_2O^+$	[HEMIM] or [C$_2$OHmim]	
$C_6H_{13}N_2^+$	[N$_{1,1}$CNC$_1$]	
$C_6H_{16}NO_3^+$	[2-HTEA]	
$C_7H_{10}N^+$	[C8py]	

Table 5.1 (continued)

Chemical formula	Abbreviation	Structure
$C_7H_{10}NO^+$	[C$_2$OHPy]	
$C_7H_{11}N_2{}^+$	[AMIM]	
$C_7H_{13}N_2{}^+$	[BIM]	
$C_7H_{13}N_2{}^+$	[PMIM] or [C$_3$mim]	
$C_7H_{13}N_2O^+$	[HOPMIM]	
$C_7H_{13}N_2O^+$	[MOEMIM] or [MOC$_2$mim]	
$C_7H_{13}N_2O^+$	[EOMMIM]	
$C_7H_{16}NO^+$	[C$_2$mmor]	
$C_7H_{16}NO_2{}^+$	[C$_2$OHmmor]	
$C_7H_{18}NO^+$	[N$_{1,1,2}$MOC$_2$]	
$C_7H_{18}NO^+$	[N$_{1,1,2}$EOC$_1$]	
$C_7H_{18}NO^+$	[N$_{1,1,2}$C$_3$OH]	
$C_8H_{12}N^+$	[E3MPy]	
$C_8H_{12}N^+$	[C$_3$py]	

Table 5.1 (continued)

Chemical formula	Abbreviation	Structure
$C_8H_{12}NO^+$	[C₃OHpy]	
$C_8H_{12}NO^+$	[EOC₁py]	
$C_8H_{12}NO^+$	[MOC₂py]	
$C_8H_{15}N_2^+$	[BMIM] or [C₄mim]	
$C_8H_{15}N_2^+$	[CNC₁mpip]	
$C_8H_{15}N_2O^+$	[EOEMIM]	
$C_8H_{15}N_2O^+$	[MOPMIM] or [MOC₃mim]	
$C_8H_{18}NO^+$	[MOC₂mpyrr]	
$C_8H_{18}NO^+$	[C₃OHmpyrr]	
$C_8H_{18}NO^+$	[C₂OHmpip]	
$C_8H_{20}N^+$	[N₁,₁,₂,₄]	
$C_8H_{20}NO^+$	[N₁,₁,₂MOC₃]	
$C_8H_{18}NO_2^+$	[MOC₂mmor]	
$C_8H_{18}NO_2^+$	[EOC₁mmor]	

Table 5.1 (continued)

Chemical formula	Abbreviation	Structure
$C_8H_{18}NO_2^+$	[EOC₃mmor]	
$C_8H_{20}N^+$	[N1124]	
$C_8H_{20}NO^+$	[N$_{1,1,2}$EOC₂]	
$C_9H_{14}N^+$	[BPy] or [C₄py]	
$C_9H_{14}NO^+$	[EOC₂py]	
$C_9H_{14}NO^+$	[MOC₃py]	
$C_9H_{15}N_2^+$	[NmmC8py]	
$C_9H_{17}N_2^+$	[PeMIM] or [C₅mim]	
$C_9H_{17}N_2^+$	[BEIM] or [C₄C₂im]	
$C_9H_{17}N_2^+$	[BDMIM]	
$C_9H_{17}N_2O_2^+$	[M(OE)₂MIM]	
$C_9H_{20}N^+$	[BMPyrr] or [C₄mpyrr]	
$C_9H_{20}NO^+$	[BMMor]	

Table 5.1 (continued)

Chemical formula	Abbreviation	Structure
$C_9H_{20}NO^+$	[EOC$_2$mpyrr]	
$C_9H_{20}NO^+$	[MOC$_3$mpyrr]	
$C_9H_{20}NO^+$	[MOC$_2$mpip]	
$C_9H_{20}NO^+$	[EOC$_1$mpip]	
$C_9H_{20}NO^+$	[C$_3$OHmpip]	
$C_9H_{20}NO^+$	[C$_4$mmor]	
$C_9H_{20}NO_2^+$	[EOC$_2$mmor]	
$C_9H_{20}NO_2^+$	[MOC$_3$mmor]	
$C_{10}H_{16}N^+$	[B3MPy] or [C$_4$mpy]	
$C_{10}H_{16}N^+$	[iC$_4$mpy]	
$C_{10}H_{16}N^+$	[C5py]	
$C_{10}H_{19}N_2^+$	[HMIM] or [C$_6$mim]	

Table 5.1 (continued)

Chemical formula	Abbreviation	Structure
$C_{10}H_{22}N^+$	[BMPip] or [C$_4$mpip]	
$C_{10}H_{22}NO^+$	[MOC$_3$mpip]	
$C_{10}H_{22}NO^+$	[EOC$_2$mpip]	
$C_{11}H_{13}N_2^+$	[Bnmim]	
$C_{11}H_{18}N^+$	[HPy] or [C$_6$py]	
$C_{11}H_{18}N^+$	[B2M3MPy]	
$C_{11}H_{18}N^+$	[B3M5MPy]	
$C_{11}H_{18}N^+$	[C$_4$mmpy]	
$C_{11}H_{18}N^+$	[C$_4$mmpy]	
$C_{11}H_{19}N_2^+$	[Py4-4NMe$_2$] or [NmmC4py]	
$C_{11}H_{21}N_2^+$	[HDMIM] or [C$_6$mmim]	
$C_{11}H_{21}N_2^+$	[HepMIM] or [C$_7$mim]	

Table 5.1 (continued)

Chemical formula	Abbreviation	Structure
$C_{11}H_{21}N_2^+$	[HEIM]	
$C_{11}H_{21}N_2O_3^+$	[M(OE)$_3$MIM]	
$C_{11}H_{22}NO^+$	[BQuinu]	
$C_{11}H_{24}N^+$	[HMPyrr] or [C$_6$mpyrr]	
$C_{11}H_{24}N^+$	[BEPip]	
$C_{12}H_{20}N^+$	[H3MPy] or [C$_6$mpy]	
$C_{12}H_{20}N^+$	[B2M3M5MPy]	
$C_{12}H_{20}N^+$	[iC$_6$mpy]	
$C_{12}H_{23}N_2^+$	[OMIM] or [C$_8$mim]	
$C_{12}H_{24}NO^+$	[BTrop]	
$C_{12}H_{26}N^+$	[HMPip]	

Table 5.1 (continued)

Chemical formula	Abbreviation	Structure
$C_{13}H_{16}N^+$	[C4qui]	
$C_{13}H_{22}N^+$	[OPy]	
$C_{13}H_{23}N_2^+$	[NmmC6py]	
$C_{13}H_{24}N^+$	[BB(CN)Pyrr]	
$C_{13}H_{25}N_2^+$	[C9MIM] or [C9mim]	
$C_{13}H_{25}N_2O_4^+$	[M(OE)4MIM]	
$C_{13}H_{28}N^+$	[OMPyrr] or [C8mpyrr]	
$C_{13}H_{28}NO^+$	[OMMor]	
$C_{13}H_{30}P^+$	[P4,4,4,1]	
$C_{13}H_{30}P^+$	[Pi(4,4,4)1]	
$C_{14}H_{24}N^+$	[OMPy]	

Table 5.1 (continued)

Chemical formula	Abbreviation	Structure
$C_{14}H_{24}N^+$	[C8mpy]	
$C_{14}H_{27}N_2{}^+$	[C10MIM] or [C_{10}mim]	
$C_{14}H_{30}N^+$	[OMPIP]	
$C_{15}H_{20}N^+$	[C6qui]	
$C_{15}H_{26}N^+$	[O2M3MPy]	
$C_{15}H_{26}N^+$	[O3M5MPy]	
$C_{15}H_{30}NO^+$	[OQuinu]	
$C_{15}H_{32}N^+$	[OEPip]	
$C_{16}H_{28}N^+$	[O2M5EPy]	
$C_{16}H_{28}N^+$	[O2M3M5MPy]	

Table 5.1 (continued)

Chemical formula	Abbreviation	Structure
$C_{16}H_{32}NO^+$	[OTrop]	
$C_{16}H_{34}N^+$	[C6C6pyrr]	
$C_{17}H_{24}N^+$	[C8qui]	
$C_{17}H_{30}N^+$	[BA]	
$C_{17}H_{31}N_2O_2^+$	[C10C(O)OEtMIM]	
$C_{18}H_{35}N_2^+$	[C14MIM]	
$C_{19}H_{34}N^+$	[CetPy]	
$C_{20}H_{39}N_2^+$	[C16MIM] or [C$_{16}$mim]	
$C_{21}H_{38}N^+$	[N$_{1,1,12}$Bn]	
$C_{22}H_{43}N_2^+$	[C18MIM]	
$C_{23}H_{42}N^+$	[N$_{1,1,14}$Bn]	

Table 5.1 (continued)

Chemical formula	Abbreviation	Structure
$C_{27}H_{42}NO_2^+$	[BE]	
$C_{27}H_{58}N_3^+$	[(di-h)2DMG]	
$C_{32}H_{68}P^+$	[P6,6,6,14]	

Table 5.2: Chemical formula, abbreviation, and structures of some common anions.

Chemical formula	Abbreviation	Structure
Cl^-	[Cl]	$Cl^{\ominus}$
Br^-	[Br]	$Br^{\ominus}$
I^-	[I]	$I^{\ominus}$
BF_4^-	[BF₄]	
HO_4S^-	[HSO₄]	
PF_6^-	[PF₆]	
$CO_3SF_3^-$	[TFO] or [TfO]	
CNS^-	[SCN]	$N{\equiv\!\!\!\equiv}\,S^{\ominus}$

Table 5.2 (continued)

Chemical formula	Abbreviation	Structure
CHO_2^-	[For]	
$CH_3O_3S^-$	[MeSO_3]	
$CH_3O_4S^-$	[MeSO_4]	
$C_2NF_6^-$	[N(CF_3)_2]	
$C_2O_2F_3^-$	[TFA]	
$C_2NO_4S_2F_6^-$	[NTF2] or [Tf_2N]	
$C_2N_3^-$	[DCA]	
$C_2H_3O_2^-$	[Ace]	
$C_2H_5O_4S^-$	[EtSO_4]	
$C_3H_5O_2^-$	[Pr]	
$C_4N_4B^-$	[B(CN)_4]	

Table 5.2 (continued)

Chemical formula	Abbreviation	Structure
$C_4H_7O_2^-$	[But]	
$C_4H_7O_2^-$	[iBut]	
$C_5H_9O_2^-$	[Pe]	
$C_5H_{11}O_6S^-$	[MDEGSO$_4$]	
$C_6F_{18}P^-$	[TPTP]	
$C_6H_5O_3S^-$	[pTS]	
$C_7H_5O_3^-$	[Sal]	
$C_7H_7O_3S^-$	[TOS] or [Tos]	
$C_8H_{15}O_2^-$	[Cap]	
$C_8H_{17}O_4S^-$	[C8OSO$_3$]	

This chapter expresses several reliable models for the assessment of the toxicity of ILs. Each approach will be illustrated in the following sections.

5.1 Toxicity of ILs Based on *Vibrio fischeri* Through the Structure of Cations with Specific Anions

Inhibition assays are the most widely used method to determine toxicological risk in an aqueous medium [349]. The most common bacterial bioassay uses *V. fischeri*, formerly known as *Photobacterium phosphoreum*, a marine gram-negative bacterium [350]. Since this inhibition assay has the most rapid, cost-effective, sensitive, and reproducible assay, it is the standard (eco)toxicological bioassay in Europe (DIN EN ISO 11348) ("ISO 11348–3. Water quality. Determination of the inhibitory effect of water samples on the light emission of *Vibrio fischeri* (Luminescent bacteria test). Part 3: Method using freeze-dried bacteria," 2007). Different luminescence inhibition tests including *V. fischeri* have been established for the analysis of aqueous samples [349]. For example, the (eco)toxicity of several ILs is determined by the Microtox® Toxicity Test, which is one of the most widely used bioassay tests because of its intense and stable light emission [351]. Moreover, this test gives the high sensitivity to different compounds and the flexibility which a marine bacterium provides [351]. This test can determine the acute toxicity of aqueous compounds by measuring decreases in light output from the luminescent bacterium *V. fischeri.* Since the light emission is directly proportional to the metabolic activity of the bacterial population, any inhibition of enzymatic activity causes a corresponding decrease in luminescence [350].

A large database including 187 ILs corresponding to 250 experimental data toxicity data on *V. fischeri* covering a diversity of 78 cations and 27 anions has been used to develop a simple model, which is given as follows [352]:

$$-\log EC_{50}(IL, \mu M) = -5.06 + [0.312n_C - 0.0797n_H + 0.319n_N - 0.501n_O]_{cation}$$

$$+ [0.0841n_H + 0.0746n_F]_{anion} + 1.0499EC^+_{50,IL} - 1.0540EC^-_{50,IL} \quad (5.1^*)$$

where n_C, n_H, n_N, and n_O in the first bracket are the number of carbon, hydrogen, nitrogen, and oxygen atoms in cation as well as n_H and n_F in the second bracket are the number of hydrogen and fluorine atoms in the anion of a desired IL, respectively; the parameters $EC^+_{50,IL}$ and $EC^-_{50,IL}$ show increasing and decreasing toxicities by structural moieties, respectively. Table 5.3 shows the values of $EC^+_{50,IL}$ and $EC^-_{50,IL}$.

As shown in Table 5.3, the length of the alkyl group attached to cations is one of the major important factors for specifying different values of $EC^+_{50,IL}$ and $EC^-_{50,IL}$. The toxicity of some pyridinium-, imidazolium-, and ammonium-based ILs toward the bioluminescent photobacterium *V. fischeri* showed that the alkyl chain length has the main influence [353].

Table 5.3: Specific cation/anion moieties for estimation of $EC_{50,IL}^{+}$ and $EC_{50,IL}^{-}$.

Cation	Anion	$EC_{50,IL}^{+}$
 $m = 1$ or 2 R containing –OH group	 R = H or alkyl	0.7
 $n = 0$ or 1 $n = 0$ or 1 or 2		1.2
 $n = 0$ or 1		0.9
 R_1, R_2, and R_3 are methyl or ethyl groups	$Cl^{\ominus}$, $Br^{\ominus}$	1.1
 R containing –C(=O)-O- group or more than one etheric functional group		1.2
 $n = 6–8$		1.0

Table 5.3 (continued)

Cation	Anion	$EC_{50,IL}^+$
R₁, R₂, and R₃ are methyl or ethyl groups		0.8
n = 7–9		1.3
n = 3–7 and R₁ and R₂ are methyl groups		0.8
R containing more than three etheric functional groups		1.0
		2.6
n = 0 or 1		0.7

Table 5.3 (continued)

Cation	Anion	$EC^+_{50,IL}$
 n = 0–3 and R containing double or triple bonds	$Cl^{\ominus}$, $Br^{\ominus}$	0.5
 n = 1–3 and R containing double or triple bonds		2.6
 n = 0–3		2.2
 n = 1–3 and R containing –OH group		1.3
 m = 0 or 1 and *n* = 0–2		1.0
 n = 0–3	$I^{\ominus}$	1.6
 n = 0–3	$N\equiv S^{\ominus}$	1.0

Table 5.3 (continued)

Cation	Anion	$EC_{50,IL}^{+}$
$n = 0–2$		1.0
$n = 0–3$ and R containing –CN group		1.0
$n = 0–5$		0.7
$n = 1–3$		1.0
$n = 1–3$		1.0
		1.5

Example 5.1: For [O2M3M5MPy][NTF2], the experimental value of $-\log EC_{50}(IL, \ \mu M)$ is −0.62. (a) Use eq. **(5.1*)** to calculate $-\log EC_{50}(IL, \ \mu M)$. (b) Calculate the deviation of eq. **(5.1*)** from experimental data.

Answer: (a) The structure of [O2M3M5MPy][NTF2] is

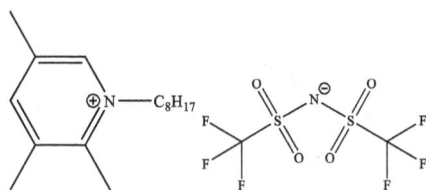

For cation, $n_C = 16$, $n_H = 28$, $n_N = 1$, and $n_O = 0$. For anion, $n_H = 0$ and $n_F = 6$. According to Table 5.3, $EC_{50,IL}^+ = 0.8$ and $EC_{50,IL}^- = 0$. The use of eq. (**5.1***) gives

$$-\log EC_{50}(IL,\ \mu M) = -5.06 + [0.312n_C - 0.0797n_H + 0.319n_N - 0.501n_O]_{\text{cation}}$$
$$+ [0.0841n_H + 0.0746n_F]_{\text{anion}} + 1.0499EC_{50,IL}^+ - 1.0540EC_{50,IL}^-$$
$$= -5.06 + 0.312(16) - 0.0797(28) + 0.319(1) - 0.501(0)$$
$$+ 0.0841(0) + 0.0746(6) + 1.0499(0.8) - 1.0540(0) = -0.69$$

(b) Dev = −0.69−(−0.62) = −0.07

5.2 Relationships of the Toxicity with the Structure and the 1-Octanol–Water Partition Coefficient of ILs

Montalban et al. [353] studied the relevance of some structural features of ILs including the cation core (tetrafluoroborate and ethyl sulfate anions), the alkyl chain length (hexafluorophosphate, tetrafluoroborate, bis(trifluoromethyl sulfonyl)imide, and chloride anions), the impact of different anions (imidazolium family), the existence of a functionalized lateral chain (chloride anion), the influence of a second short alkyl chain (bis(trifluoromethyl sulfonyl) imide and chloride anions), and the effect of a double bond in the alkyl chain (chloride anion) on the toxicity toward the marine luminescent bacterium *V. fischeri*. They found that there is a linear relationship between −log EC_{50} versus alkyl chain length as follows:

$$-\log EC_{50}(IL,\ \mu M) = -5.22 + 0.52n C_{R1} \qquad (5.2)$$

All types of anions showed increased IL toxicity as the alkyl chain length increased:
[OMIM][PF$_6$] > [HEIM][PF$_6$] > [BMIM][PF$_6$] > [EMIM][PF$_6$];
[OMIM][BF$_4$] > [BMIM][BF$_4$];
[OMIM][NTF2] > [HEIM][NTF2] > [BMIM][NTF2] > [EMIM][NTF2];
[OMIM][Cl] > [HEIM][Cl] > [BMIM][Cl] > [PMIM][Cl] > [EMIM][Cl]

Montalban et al. [353] compared the toxicity using [BMIM] and [EMIM] cations with different types of anions. They established the following trend in IL toxicity:

[NTF2] > [PF6] > [BF4] ≈ [MeSO$_4$] ≈ [MDEGSO$_4$] > [TFO] > [Ace] > [EtSO$_4$] > [Cl]

The most toxic anions are those containing fluorine atoms in their structure and their toxicity increases with an increment in the number of fluorine atoms [354]. Thus, the [NTF2] anion with six fluorine atoms was the most toxic toward *V. fischeri*. Fluoride ions can act as enzymatic poisons, inhibiting enzyme activity and ultimately interrupting metabolic processes, such as glycolysis and the synthesis of proteins [355].

Since lipophilicity and hydrophobicity of ILs are linked to aquatic toxicity, Montalban et al. [353] deduced a relationship between IL hydrophobicity in terms of log K_{OW} and $-\log EC_{50}$ as follows:

$$-\log EC_{50}(IL, \ \mu M) = -1.436 + 0.859 \log K_{OW} - 0.119(\log K_{OW})^2 \qquad (5.3^*)$$

The correlation may be quite poor for some of the ILs studied. Since the EC_{50} values are of the same order as volatile organic compounds, it can be concluded that the aquatic toxicity of these ILs is quite similar to that of industrial volatile organic compounds and will not be accumulated in the environment. Moreover, the replacement of a conventional process with a process based on ILs as new solvents may be environmentally attractive if the ILs permit a high degree of reusability due to their easy separation.

Example 5.2: The experimental values of log K_{OW} and $-\log EC_{50}$ for [BMIM][PF$_6$] are -1.4908 [356] and -3.07 [357], respectively. (a) Use eq. **(5.3*)** to calculate $-\log EC_{50}(IL, \mu M)$. (b) Calculate the deviation of eq. **(5.3*)** from experimental data.

Answer: (a) The use of eq. **(5.3*)** gives

$$-\log EC_{50}(IL, \ \mu M) = -1.436 + 0.859 \log K_{OW} - 0.119(\log K_{OW})^2$$
$$= -1.436 + 0.859(-1.4908) - 0.119(-1.4908)^2 = -2.981$$

(b) Dev $= -2.981 - (-3.05) = 0.07$

5.3 Using a Simple Group Contribution Method for Some ILs

Luis et al. [345] introduced a simple group contribution (GC) method for the prediction of toxicity of some ILs toward *V. fischeri*. This method can be applied to several categories of ILs including imidazolium, pyridinium, and pyrrolidinium-based ILs with anions [Cl], [Br], [BF$_4$], [PF$_6$], [MeSO$_4$], [EtSO$_4$], and [DCA]. Structural representations of ILs are given in Figure 5.1.

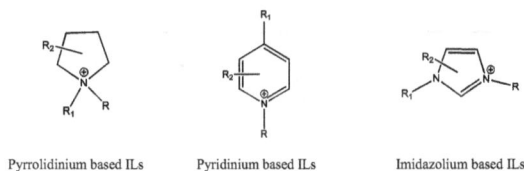

Pyrrolidinium based ILs   Pyridinium based ILs   Imidazolium based ILs

Figure 5.1: Structural representations of cations used in the group contribution method.

Luis et al. [345] introduced eq. **(5.4*)** for the calculation of $-\log EC_{50}$ toward *V. fischeri* as follows:

$$-\log EC_{50}(IL,\ \mu M) = -4.76 + 4.94\left(\sum_i a_i Ani_i + \sum_j c_j Cat_j + \sum_k s_k Sub_k\right) \qquad (5.4^*)$$

where Ani_i, Cat_j, and Sub_k are the molecular descriptors for ILs; a_i, c_j, and s_k are the contributions of each group to the toxicity, and the summation is taken over all groups. These variables have a nonzero value when the group is present in the molecule. Table 5.4 shows GCs to eq. **(5.4*)**.

Table 5.4: Group contributions to eq. **(5.4*)**.

Group	Molecular descriptor	Comments	Contribution
Anion (Ani)	Ani_1	Influence of anions: tetrafluoroborate ([BF$_4$]). Value = 1 if it exists and 0 if not.	−0.577
	Ani_2	Influence of anions: hexafluorophosphate ([PF$_6$]), chloride ([Cl]), and methylsulfate ([MeSO$_4$]). Value = 1 if it exists and 0 if it does not.	−0.510
	Ani_3	Influence of anions: bromide ([Br]), dicyanamide ([DCA]), and ethylsulfate ([EtSO$_4$]). Value = 1 if it exists and 0 if not.	−0.440
Cation (Cat)	Imida	Influence of imidazolium cation. Value = 1 if it exists and 0 if not.	0.197
	Pyrid Pyrrol	Influence of pyridinium cation. Value = 1 if it exists and 0 if not. Influence of pyrrolidinium cation. Value = 1 if it exists and 0 if not.	0.331 0.028
Substitution (Sub)	R	Influence of the number of carbons in long-chain (R: 0 to 10).	0.110
	R_1	Influence of the number of carbons in the short chain (R_1: 1, 2)	0.114
	R_2	Influence of additional short chain in the molecule (R_2: 0, 1, 2).	0.067

Example 5.3. Use eq. **(5.4*)** to calculate $-\log EC_{50}$ for [EMIM][Cl].
Answer: According to Table 5.4, the values of $A_2 = 1$, Imida = 1, $R = 2$, $R_1 = 1$, and $R_2 = 1$. The use of these molecular descriptors in eq. **(5.4*)** gives

$$-\log EC_{50}(IL,\ \mu M) = -4.76 + 4.94\left(\sum_i a_i Ani_i + \sum_j c_j Cat_j + \sum_k s_k Sub_k\right)$$

$$= -4.76 + 4.94((-0.510)(1) + (0.197)(1) + (0.110)(2) + (0.114)(1) + (0.067)(1))$$

$$= -4.33$$

The calculated value of $-\log EC_{50}$ is close to the experimental value (−4.55 [345]).

5.4 Using Atomic Electrostatic Potential Descriptors for Predicting the Ecotoxicity of ILs Toward Leukemia Rat Cell Line (ICP-81)

Kang et al. [358] introduced the non-integer GC (NGC) method based on the atomic electrostatic potential descriptors including the average values of electrostatic potential (AV_{EP}) and the electrostatic potential surface area (S_{EP}) of atoms in the cations and anions of ILs for modeling. The NGC-2 model based on S_{EP} descriptors exhibits better predictability than NGC-1 based on AV_{EP} descriptors. Since the NGC-2 model also shows better performance than the traditional GC method, it has high potential in terms of generalization and applicability. The traditional GC and NGC-2 methods for assessing the toxicity of ILs toward ICP-81 are given as follows:

$$-\log EC_{50}(IL, \mu M) = \beta_0 + \sum_{m=1}^{26} \beta_m X_m \qquad (5.5^*)$$

where m represents the mth group in an IL; X_m is the number of the corresponding group occurring in an IL, including its cation and anion; β_m is the regressed contribution of the corresponding group; β_0 is a constant. Table 5.5 gives the values of β_0 to β_{26} corresponding to the mth group.

Table 5.5: The values of β_0 to β_{26} corresponding to the mth group for the traditional GC and NGC-2 methods.

Group type	Structures	Coefficients of eq. (5.5*) by the traditional GC method	Coefficients of eq. (5.5*) by the NGC-2 model
	Cation	$\beta_0 = -5.374$	$\beta_0 = -16.861$
A		$\beta_1 = -0.084$	$\beta_1 = -0.118$
B		$\beta_2 = 0.215$	$\beta_2 = 0$
C		$\beta_3 = 0$	$\beta_3 = -0.055$

Table 5.5 (continued)

Group type	Structures	Coefficients of eq. (5.5*) by the traditional GC method	Coefficients of eq. (5.5*) by the NGC-2 model
D		$\beta_4 = 0.381$	$\beta_4 = 0$
E	$N(CH_2)_2(CH_3)_2$	$\beta_5 = -0.181$	$\beta_5 = -0.212$
F		$\beta_6 = 1.655$	$\beta_6 = 1.554$
G		$\beta_7 = 0.134$	$\beta_7 = -0.183$
	Anion		
H	[TPTP]	$\beta_8 = 2.779$	$\beta_8 = 14.158$
I	[BF$_4$]	$\beta_9 = 1.846$	$\beta_9 = 13.249$
J	[PF$_6$]	$\beta_{10} = 1.604$	$\beta_{10} = 12.986$
K	[Cl]	$\beta_{11} = 1.569$	$\beta_{11} = 12.923$
L	[Br]	$\beta_{12} = 1.373$	$\beta_{12} = 12.753$
M	[I]	$\beta_{13} = 1.385$	$\beta_{13} = 12.809$
N	[SCN]	$\beta_{14} = 1.446$	$\beta_{14} = 12.827$
	Substituents		
O	-SO$_3$	$\beta_{15} = 1.697$	$\beta_{15} = 12.925$
P	-S(=O)$_2$-	$\beta_{16} = 0.365$	$\beta_{16} = 3.301$
Q	-COO-	$\beta_{17} = 1.568$	$\beta_{17} = 12.767$
R	-CF$_3$	$\beta_{18} = -0.113$	$\beta_{18} = 0.036$
S	-CH$_3$	$\beta_{19} = -0.048$	$\beta_{19} = -0.040$
T	-CH$_2$-	$\beta_{20} = 0.321$	$\beta_{20} = 0.325$
U	-H(ring)	$\beta_{21} = -0.119$	$\beta_{21} = -0.081$
V	-O- or [-O]$^-$	$\beta_{22} = -0.057$	$\beta_{22} = 0.017$

Table 5.5 (continued)

Group type	Structures	Coefficients of eq. (5.5*) by the traditional GC method	Coefficients of eq. (5.5*) by the NGC-2 model
W	-OH	$\beta_{23} = -0.253$	$\beta_{23} = -0.406$
X		$\beta_{24} = 0.817$	$\beta_{24} = 1.140$
Y	-N- or >N-	$\beta_{25} = 1.508$	$\beta_{25} = 6.422$
Z	-CN	$\beta_{26} = -0.05$	$\beta_{26} = -0.098$

Table 5.6 presents the X_m values of 26 groups for 74 cations and 15 anions in 140 ILs for the traditional GC model.

Table 5.6: The X_m values of 26 groups for 74 cations and 15 anions for the traditional GC model.

Group type / Abbreviation of cations/anions	A	B	C	D	E	F	G	H	I	J	K	L	M	N	O	P	Q	R	S	T	U	V	W	X	Y	Z
Cations																										
[EOC2mpyrr]	0	1	0	0	0	0	0	0	0	0	0	0	0	0	0	0	0	2	3	8	1	0	0	0	0	0
[C2OHmim]	1	0	0	0	0	0	0	0	0	0	0	0	0	0	0	0	0	1	2	3	0	1	0	0	0	0
[MOC2mpyrr]	0	1	0	0	0	0	0	0	0	0	0	0	0	0	0	0	0	2	2	8	1	0	0	0	0	0
[MOC2mim]	1	0	0	0	0	0	0	0	0	0	0	0	0	0	0	0	0	2	2	3	1	0	0	0	0	0
[MOC3mpip]	0	0	0	1	0	0	0	0	0	0	0	0	0	0	0	0	0	2	3	10	1	0	0	0	0	0
[Bnmim]	1	0	0	0	0	0	0	0	0	0	0	0	0	0	0	0	0	1	1	3	0	0	1	0	0	0
[C4mpip]	0	0	0	1	0	0	0	0	0	0	0	0	0	0	0	0	0	2	3	10	0	0	0	0	0	0
[C4mpyrr]	0	1	0	0	0	0	0	0	0	0	0	0	0	0	0	0	0	2	3	8	0	0	0	0	0	0
[C4mmpy]	0	0	1	0	0	0	0	0	0	0	0	0	0	0	0	0	0	3	3	3	0	0	0	0	0	0
[C4C2im]	1	0	0	0	0	0	0	0	0	0	0	0	0	0	0	0	0	2	4	3	0	0	0	0	0	0
[C4mim]	1	0	0	0	0	0	0	0	0	0	0	0	0	0	0	0	0	2	3	3	0	0	0	0	0	0
[C4mpy]	0	0	1	0	0	0	0	0	0	0	0	0	0	0	0	0	0	2	3	4	0	0	0	0	0	0
[NmmC4py]	0	0	1	0	0	0	0	0	0	0	0	0	0	0	0	0	0	3	3	4	0	0	0	0	1	0
[C4py]	0	0	1	0	0	0	0	0	0	0	0	0	0	0	0	0	0	1	3	5	0	0	0	0	0	0
[C10mim]	1	0	0	0	0	0	0	0	0	0	0	0	0	0	0	0	0	2	9	3	0	0	0	0	0	0
[C2mim]	1	0	0	0	0	0	0	0	0	0	0	0	0	0	0	0	0	2	1	3	0	0	0	0	0	0
[C7mim]	1	0	0	0	0	0	0	0	0	0	0	0	0	0	0	0	0	2	6	3	0	0	0	0	0	0
[C16mim]	1	0	0	0	0	0	0	0	0	0	0	0	0	0	0	0	0	2	15	3	0	0	0	0	0	0
[C6mpyrr]	0	1	0	0	0	0	0	0	0	0	0	0	0	0	0	0	0	2	5	8	0	0	0	0	0	0
[C6mmim]	1	0	0	0	0	0	0	0	0	0	0	0	0	0	0	0	0	3	5	2	0	0	0	0	0	0
[C6mim]	1	0	0	0	0	0	0	0	0	0	0	0	0	0	0	0	0	2	5	3	0	0	0	0	0	0

Table 5.6 (continued)

Group type / Abbreviation of cations/anions	A	B	C	D	E	F	G	H	I	J	K	L	M	N	O	P	Q	R	S	T	U	V	W	X	Y	Z
[C₆mpy]	0	0	1	0	0	0	0	0	0	0	0	0	0	0	0	0	0	0	2	5	4	0	0	0	0	0
[iC₆mpy]	0	0	1	0	0	0	0	0	0	0	0	0	0	0	0	0	0	0	2	5	4	0	0	0	0	0
[C₆py]	0	0	1	0	0	0	0	0	0	0	0	0	0	0	0	0	0	0	1	5	5	0	0	0	0	0
[C₉mim]	1	0	0	0	0	0	0	0	0	0	0	0	0	0	0	0	0	0	2	8	3	0	0	0	0	0
[C₈mpyrr]	0	1	0	0	0	0	0	0	0	0	0	0	0	0	0	0	0	0	2	7	8	0	0	0	0	0
[C₈mim]	1	0	0	0	0	0	0	0	0	0	0	0	0	0	0	0	0	0	2	7	3	0	0	0	0	0
[C₈qui]	0	0	0	0	0	1	0	0	0	0	0	0	0	0	0	0	0	0	1	7	7	0	0	0	0	0
[C₅mim]	1	0	0	0	0	0	0	0	0	0	0	0	0	0	0	0	0	0	2	4	3	0	0	0	0	0
[C₃mim]	1	0	0	0	0	0	0	0	0	0	0	0	0	0	0	0	0	0	2	2	3	0	0	0	0	0
[C₈mpy]	0	0	1	0	0	0	0	0	0	0	0	0	0	0	0	0	0	0	2	7	4	0	0	0	0	0
[iC₄mpy]	0	0	1	0	0	0	0	0	0	0	0	0	0	0	0	0	0	0	2	3	4	0	0	0	0	0
[C₃OHpy]	0	0	1	0	0	0	0	0	0	0	0	0	0	0	0	0	0	0	0	3	5	0	1	0	0	0
[EOC₂mpip]	0	0	0	1	0	0	0	0	0	0	0	0	0	0	0	0	0	0	2	3	10	1	0	0	0	0
[EOC₂py]	0	0	1	0	0	0	0	0	0	0	0	0	0	0	0	0	0	0	1	3	5	1	0	0	0	0
[C₂OHmpip]	0	0	0	1	0	0	0	0	0	0	0	0	0	0	0	0	0	0	1	2	10	0	1	0	0	0
[C₂OHPy]	0	0	1	0	0	0	0	0	0	0	0	0	0	0	0	0	0	0	0	2	5	0	1	0	0	0
[MOC₂mpip]	0	0	0	1	0	0	0	0	0	0	0	0	0	0	0	0	0	0	2	2	10	1	0	0	0	0
[C₃OHmpyrr]	0	1	0	0	0	0	0	0	0	0	0	0	0	0	0	0	0	0	1	3	8	0	1	0	0	0
[MOC₃py]	0	0	1	0	0	0	0	0	0	0	0	0	0	0	0	0	0	0	1	3	5	1	0	0	0	0
[C₄qui]	0	0	0	0	0	1	0	0	0	0	0	0	0	0	0	0	0	0	1	3	7	0	0	0	0	0
[CNC₁mpip]	0	0	0	1	0	0	0	0	0	0	0	0	0	0	0	0	0	0	1	1	10	0	0	0	0	1
[EOC₁mpip]	0	0	0	1	0	0	0	0	0	0	0	0	0	0	0	0	0	0	2	2	10	1	0	0	0	0
[EOC₁py]	0	0	1	0	0	0	0	0	0	0	0	0	0	0	0	0	0	0	1	2	5	1	0	0	0	0
[C₅py]	0	0	1	0	0	0	0	0	0	0	0	0	0	0	0	0	0	0	1	4	5	0	0	0	0	0
[C₃py]	0	0	1	0	0	0	0	0	0	0	0	0	0	0	0	0	0	0	1	2	5	0	0	0	0	0
[N₁,₁,₂EOC₂]	0	0	0	0	1	0	0	0	0	0	0	0	0	0	0	0	0	0	2	2	0	1	0	0	0	0
[EOC₂mmor]	0	0	0	0	0	0	1	0	0	0	0	0	0	0	0	0	0	0	2	3	8	1	0	0	0	0
[MOC₂mmor]	0	0	0	0	0	0	1	0	0	0	0	0	0	0	0	0	0	0	2	2	8	1	0	0	0	0
[EOC₃mmor]	0	0	0	0	0	0	1	0	0	0	0	0	0	0	0	0	0	0	1	3	8	0	1	0	0	0
[MOC₃mmor]	0	0	0	0	0	0	1	0	0	0	0	0	0	0	0	0	0	0	2	3	8	1	0	0	0	0
[C₄mmor]	0	0	0	0	0	0	1	0	0	0	0	0	0	0	0	0	0	0	2	3	8	0	0	0	0	0
[EOC₁mmor]	0	0	0	0	0	0	1	0	0	0	0	0	0	0	0	0	0	0	2	2	8	1	0	0	0	0
[C₂mmor]	0	0	0	0	0	0	1	0	0	0	0	0	0	0	0	0	0	0	2	1	8	0	0	0	0	0
[N₁,₁CNC₁]	0	0	0	0	1	0	0	0	0	0	0	0	0	0	0	0	0	0	1	0	0	0	0	0	0	1
[N₁,₁,₁₄Bn]	0	0	0	0	1	0	0	0	0	0	0	0	0	0	0	0	0	0	1	12	5	0	0	1	0	0
[N₁,₁,₂MOC₃]	0	0	0	0	1	0	0	0	0	0	0	0	0	0	0	0	0	0	2	2	0	1	0	0	0	0
[C₄mmpy]	0	0	1	0	0	0	0	0	0	0	0	0	0	0	0	0	0	0	3	3	3	0	0	0	0	0
[C₈py]	0	0	1	0	0	0	0	0	0	0	0	0	0	0	0	0	0	0	1	1	5	0	0	0	0	0
[NmmC₆py]	0	0	1	0	0	0	0	0	0	0	0	0	0	0	0	0	0	0	3	5	4	0	0	0	1	0
[NmmC₈py]	0	0	1	0	0	0	0	0	0	0	0	0	0	0	0	0	0	0	3	1	4	0	0	0	1	0
[N₁,₁,₂MOC₂]	0	0	0	0	1	0	0	0	0	0	0	0	0	0	0	0	0	0	2	1	0	1	0	0	0	0
[C₆C₆pyrr]	0	1	0	0	0	0	0	0	0	0	0	0	0	0	0	0	0	0	2	10	8	0	0	0	0	0
[MOC₂py]	0	0	1	0	0	0	0	0	0	0	0	0	0	0	0	0	0	0	1	2	5	1	0	0	0	0
[C₃OHmpip]	0	0	0	1	0	0	0	0	0	0	0	0	0	0	0	0	0	0	1	3	10	0	1	0	0	0

Table 5.6 (continued)

Group type / Abbreviation of cations/anions	A	B	C	D	E	F	G	H	I	J	K	L	M	N	O	P	Q	R	S	T	U	V	W	X	Y	Z
[MOC$_3$mpyrr]	0	1	0	0	0	0	0	0	0	0	0	0	0	0	0	0	0	0	2	3	8	1	0	0	0	0
[C$_6$qui]	0	0	0	0	0	1	0	0	0	0	0	0	0	0	0	0	0	1	5	7	0	0	0	0	0	0
[MOC$_3$mim]	1	0	0	0	0	0	0	0	0	0	0	0	0	0	0	0	0	0	2	3	3	1	0	0	0	0
[CNC$_1$mim]	1	0	0	0	0	0	0	0	0	0	0	0	0	0	0	0	0	1	1	3	0	0	0	0	0	1
[C$_2$OHmmor]	0	0	0	0	0	0	1	0	0	0	0	0	0	0	0	0	0	1	2	8	1	0	0	0	0	0
[N$_{1,1,12}$Bn]	0	0	0	0	1	0	0	0	0	0	0	0	0	0	0	0	0	1	10	5	0	0	1	0	0	0
[N$_{1,1,2}$EOC$_1$]	0	0	0	0	1	0	0	0	0	0	0	0	0	0	0	0	0	2	1	0	1	0	0	0	0	0
[N$_{1,1,2}$C$_3$OH]	0	0	0	0	1	0	0	0	0	0	0	0	0	0	0	0	0	1	2	0	0	1	0	0	0	0
[N$_{1,1,2,4}$]	0	0	0	0	1	0	0	0	0	0	0	0	0	0	0	0	0	2	2	0	0	0	0	0	0	0
Anions																										
[BF$_4$]	0	0	0	0	0	0	0	0	1	0	0	0	0	0	0	0	0	0	0	0	0	0	0	0	0	0
[Br]	0	0	0	0	0	0	0	0	0	0	0	1	0	0	0	0	0	0	0	0	0	0	0	0	0	0
[Cl]	0	0	0	0	0	0	0	0	0	0	1	0	0	0	0	0	0	0	0	0	0	0	0	0	0	0
[DCA]	0	0	0	0	0	0	0	0	0	0	0	0	0	0	0	0	0	0	0	0	0	0	0	0	1	2
[HSO$_4$]	0	0	0	0	0	0	0	0	0	0	0	0	0	0	1	0	0	0	0	0	0	1	0	0	0	0
[I]	0	0	0	0	0	0	0	0	0	0	0	0	1	0	0	0	0	0	0	0	0	0	0	0	0	0
[MeSO$_3$]	0	0	0	0	0	0	0	0	0	0	0	0	0	0	1	0	0	0	1	0	0	0	0	0	0	0
[MeSO$_4$]	0	0	0	0	0	0	0	0	0	0	0	0	0	0	1	0	0	0	1	0	0	1	0	0	0	0
[PF$_6$]	0	0	0	0	0	0	0	0	0	1	0	0	0	0	0	0	0	0	0	0	0	0	0	0	0	0
[Tf$_2$N]	0	0	0	0	0	0	0	0	0	0	0	0	0	0	2	0	2	0	0	0	0	0	0	1	0	0
[SCN]	0	0	0	0	0	0	0	0	0	0	0	0	0	1	0	0	0	0	0	0	0	0	0	0	0	0
[Tos]	0	0	0	0	0	0	0	0	0	0	0	0	0	0	1	0	0	0	1	0	4	0	0	1	0	0
[TFA]	0	0	0	0	0	0	0	0	0	0	0	0	0	0	0	1	1	0	0	0	0	0	0	0	0	0
[TfO]	0	0	0	0	0	0	0	0	0	0	0	0	0	0	1	0	0	1	0	0	0	0	0	0	0	0
[TPTP]	0	0	0	0	0	0	0	1	0	0	0	0	0	0	0	0	0	0	0	0	0	0	0	0	0	0

For the proposed NGC-2 method, the descriptor (X_m) in eq. **(5.5*)** was replaced by the ratio of a specific property value to the average property value of the group in a specified data set. The values of S_{EP} of atoms in the cations and anions of ILs were calculated and used to obtain the group descriptors as follows [358]:

$$X_m = \frac{S_{EP-i,m}}{\sum_{i=1}^{N_p} S_{EP-i,m}/N} \qquad (5.6)$$

where i notes the ith IL in the data set, while m means the mth group in the IL; N is the number of each group appearing in the whole data set, and N_p is the number of ILs in the corresponding data set. Table 5.7 presents the X_m values of 26 groups for 74 cations and 15 anions in 140 ILs for the NGC-2 model.

Table 5.7: The X_m values of 26 groups for 74 cations and 15 anions for the NGC-2 model.

Group type / Abbreviation of cations/anions	A	B	C	D	E	F	G	H	I	J	K	L	M	N	O	P	Q	R	S	T	U	V	W	X	Y	Z
													Cations													
[EOC$_2$mpyrr]	0.00	0.00	0.00	0.00	0.00	0.00	0.00	0.00	0.00	0.00	0.00	0.00	0.00	0.00	0.00	0.00	0.00	0.00	1.82	2.93	6.78	0.45	0.00	0.00	0.00	0.00
[C$_2$OHmim]	1.01	0.00	0.00	0.00	0.00	0.00	0.00	0.00	0.00	0.00	0.00	0.00	0.00	0.00	0.00	0.00	0.00	0.00	1.07	2.36	3.42	0.00	0.94	0.00	0.00	0.00
[MOC$_2$mpyrr]	0.00	0.00	0.00	0.00	0.00	0.00	0.00	0.00	0.00	0.00	0.00	0.00	0.00	0.00	0.00	0.00	0.00	0.00	1.82	1.82	6.80	0.72	0.00	0.00	0.00	0.00
[MOC$_2$mim]	1.01	0.00	0.00	0.00	0.00	0.00	0.00	0.00	0.00	0.00	0.00	0.00	0.00	0.00	0.00	0.00	0.00	0.00	2.15	2.25	3.43	0.90	0.00	0.00	0.00	0.00
[MOC$_3$mpip]	0.00	0.00	0.00	0.00	0.00	0.00	0.00	0.00	0.00	0.00	0.00	0.00	0.00	0.00	0.00	0.00	0.00	0.00	1.80	2.69	8.04	1.18	0.00	0.00	0.00	0.00
[Bnmim]	1.01	0.00	0.00	0.00	0.00	0.00	0.00	0.00	0.00	0.00	0.00	0.00	0.00	0.00	0.00	0.00	0.00	0.00	1.07	1.11	9.60	0.00	0.00	0.95	0.00	0.00
[C$_4$mpip]	0.00	0.00	0.00	0.00	0.00	0.00	0.00	0.00	0.00	0.00	0.00	0.00	0.00	0.00	0.00	0.00	0.00	0.00	1.74	2.60	8.05	0.00	0.00	0.00	0.00	0.00
[C$_4$mpyrr]	0.00	0.00	0.00	0.00	0.00	0.00	0.00	0.00	0.00	0.00	0.00	0.00	0.00	0.00	0.00	0.00	0.00	0.00	1.78	2.65	6.80	0.00	0.00	0.00	0.00	0.00
[C$_4$mmpy]	0.00	0.00	0.95	0.00	0.00	0.00	0.00	0.00	0.00	0.00	0.00	0.00	0.00	0.00	0.00	0.00	0.00	0.00	2.99	3.06	3.01	0.00	0.00	0.00	0.00	0.00
[C$_4$C$_2$im]	0.93	0.00	0.00	0.00	0.00	0.00	0.00	0.00	0.00	0.00	0.00	0.00	0.00	0.00	0.00	0.00	0.00	0.00	2.08	4.26	3.36	0.00	0.00	0.00	0.00	0.00
[C$_4$mim]	1.00	0.00	0.00	0.00	0.00	0.00	0.00	0.00	0.00	0.00	0.00	0.00	0.00	0.00	0.00	0.00	0.00	0.00	2.10	3.14	3.46	0.00	0.00	0.00	0.00	0.00
[C$_4$mpy]	0.00	0.00	0.98	0.00	0.00	0.00	0.00	0.00	0.00	0.00	0.00	0.00	0.00	0.00	0.00	0.00	0.00	0.00	2.10	3.05	4.27	0.00	0.00	0.00	0.00	0.00
[NmmC$_4$py]	0.00	0.00	0.92	0.00	0.00	0.00	0.00	0.00	0.00	0.00	0.00	0.00	0.00	0.00	0.00	0.00	0.00	0.00	2.97	3.06	3.73	0.00	0.00	0.00	0.00	0.23
[C$_4$py]	0.00	0.00	1.02	0.00	0.00	0.00	0.00	0.00	0.00	0.00	0.00	0.00	0.00	0.00	0.00	0.00	0.00	0.00	1.03	3.05	5.73	0.00	0.00	0.00	0.00	0.00
[C$_{10}$mim]	1.00	0.00	0.00	0.00	0.00	0.00	0.00	0.00	0.00	0.00	0.00	0.00	0.00	0.00	0.00	0.00	0.00	0.00	2.11	9.02	3.46	0.00	0.00	0.00	0.00	0.00
[C$_2$mim]	1.01	0.00	0.00	0.00	0.00	0.00	0.00	0.00	0.00	0.00	0.00	0.00	0.00	0.00	0.00	0.00	0.00	0.00	2.11	1.13	3.47	0.00	0.00	0.00	0.00	0.00
[C$_7$mim]	1.00	0.00	0.00	0.00	0.00	0.00	0.00	0.00	0.00	0.00	0.00	0.00	0.00	0.00	0.00	0.00	0.00	0.00	2.10	6.08	3.47	0.00	0.00	0.00	0.00	0.00
[C$_{16}$mim]	1.00	0.00	0.00	0.00	0.00	0.00	0.00	0.00	0.00	0.00	0.00	0.00	0.00	0.00	0.00	0.00	0.00	0.00	2.11	14.9	3.46	0.00	0.00	0.00	0.00	0.00
[C$_6$mpyrr]	0.00	0.00	0.00	0.00	0.00	0.00	0.00	0.00	0.00	0.00	0.00	0.00	0.00	0.00	0.00	0.00	0.00	0.00	1.79	4.61	6.79	0.00	0.00	0.00	0.00	0.00
[C$_6$mmim]	0.96	0.00	0.00	0.00	0.00	0.00	0.00	0.00	0.00	0.00	0.00	0.00	0.00	0.00	0.00	0.00	0.00	0.00	2.94	4.96	2.41	0.00	0.00	0.00	0.00	0.00
[C$_6$mim]	1.00	0.00	0.00	0.00	0.00	0.00	0.00	0.00	0.00	0.00	0.00	0.00	0.00	0.00	0.00	0.00	0.00	0.00	2.11	5.10	3.46	0.00	0.00	0.00	0.00	0.00

[C₆mpy]	0.00	0.00	0.00	0.00	0.00	0.00	0.00	0.00	0.00	0.00	0.00	0.00	2.10	5.01	4.28	0.00	0.00	0.00	0.00
[iC₆mpy]	0.00	0.98	0.98	0.00	0.00	0.00	0.00	0.00	0.00	0.00	0.00	0.00	2.10	5.01	4.29	0.00	0.00	0.00	0.00
[C₆py]	0.00	0.00	1.02	0.00	0.00	0.00	0.00	0.00	0.00	0.00	0.00	0.00	1.04	5.01	5.73	0.00	0.00	0.00	0.00
[C₉mim]	1.00	0.00	0.00	0.00	0.00	0.00	0.00	0.00	0.00	0.00	0.00	0.00	2.11	8.04	3.47	0.00	0.00	0.00	0.00
[C₈mpyrr]	0.00	0.00	0.00	0.00	0.00	0.00	0.00	0.00	0.00	0.00	0.00	0.00	1.79	6.57	6.80	0.00	0.00	0.00	0.00
[C₈mim]	1.00	0.00	0.00	0.00	0.00	0.00	0.00	0.00	0.00	0.00	0.00	0.00	2.11	7.06	3.46	0.00	0.00	0.00	0.00
[C₈qui]	0.00	0.00	0.00	0.00	0.00	0.00	0.00	0.00	0.00	0.00	0.00	0.00	1.04	6.84	7.83	0.00	0.00	0.00	0.00
[C₅mim]	1.01	0.00	0.00	0.00	0.00	0.00	0.00	0.00	0.00	0.00	0.00	0.00	2.10	4.12	3.46	0.00	0.00	0.00	0.00
[C₃mim]	1.00	0.00	0.00	0.00	0.00	0.00	0.00	0.00	0.00	0.00	0.00	0.00	2.09	2.16	3.47	0.00	0.00	0.00	0.00
[C₈mpy]	0.00	0.00	0.98	0.00	0.00	0.00	0.00	0.00	0.00	0.00	0.00	0.00	2.11	6.97	4.28	0.00	0.00	0.00	0.00
[iC₄mpy]	0.00	0.00	0.98	0.00	0.00	0.00	0.00	0.00	0.00	0.00	0.00	0.00	2.10	3.06	4.28	0.00	0.00	0.00	0.00
[C₃OHpy]	0.00	0.00	1.02	0.00	0.00	0.00	0.00	0.00	0.00	0.00	0.00	0.00	0.00	3.24	5.74	0.00	1.01	0.00	0.00
[EOC₂mpip]	0.00	0.00	0.00	0.00	0.00	0.00	0.00	0.00	0.00	0.00	0.00	0.00	1.78	2.89	8.03	0.46	0.00	0.00	0.00
[EOC₂py]	0.00	0.00	1.02	0.00	0.00	0.00	0.00	0.00	0.00	0.00	0.00	0.00	1.09	3.26	5.71	0.65	0.00	0.00	0.00
[C₂OHmpip]	0.00	0.00	0.00	0.00	0.00	0.00	0.00	0.00	0.00	0.00	0.00	0.00	0.70	1.90	8.05	0.00	0.89	0.00	0.00
[C₂OHpy]	0.00	0.00	1.03	0.00	0.00	0.00	0.00	0.00	0.00	0.00	0.00	0.00	0.00	2.26	5.69	0.00	0.97	0.00	0.00
[MOC₂mpip]	0.00	0.00	0.00	0.00	0.00	0.00	0.00	0.00	0.00	0.00	0.00	0.00	1.78	1.79	8.05	0.72	0.00	0.00	0.00
[C₃OHmpyrr]	0.00	0.00	0.00	0.00	0.00	0.00	0.00	0.00	0.00	0.00	0.00	0.00	0.75	2.84	6.80	0.00	1.01	0.00	0.00
[MOC₃py]	0.00	0.00	1.03	0.00	0.00	0.00	0.00	0.00	0.00	0.00	0.00	0.00	1.08	3.12	5.74	1.19	0.00	0.00	0.00
[C₄qui]	0.00	0.00	0.00	0.00	0.00	0.00	0.00	0.00	0.00	0.00	0.00	0.00	1.03	2.93	7.82	0.00	0.00	0.00	0.00
[CNC₁mpip]	0.00	0.00	0.00	0.00	0.00	0.00	0.00	0.00	0.00	0.00	0.00	0.00	0.76	0.97	8.19	0.00	0.00	0.00	0.81
[EOC₁mpip]	0.00	0.00	0.00	0.00	0.00	0.00	0.00	0.00	0.00	0.00	0.00	0.00	1.85	1.94	8.20	0.46	0.00	0.00	0.00
[EOC₁py]	0.00	0.00	0.99	0.00	0.00	0.00	0.00	0.00	0.00	0.00	0.00	0.00	1.08	2.22	5.84	1.08	0.00	0.00	0.00
[C₅py]	0.00	0.00	1.02	0.00	0.00	0.00	0.00	0.00	0.00	0.00	0.00	0.00	1.03	4.04	5.74	0.00	0.00	0.00	0.00
[C₃py]	0.00	0.00	1.02	0.00	0.00	0.00	0.00	0.00	0.00	0.00	0.00	0.00	1.02	2.08	5.73	0.00	0.00	0.00	0.00
[N₁,₁,₂EOC₂]	0.00	0.00	0.00	0.00	0.00	0.00	0.00	0.00	1.00	0.00	0.00	0.00	1.95	1.99	0.00	0.48	0.00	0.00	0.00
[EOC₂mmor]	0.00	0.00	0.00	0.00	0.00	0.00	0.00	0.00	1.00	0.00	0.00	0.00	1.79	2.90	6.55	0.45	0.00	0.00	0.00
[MOC₂mmor]	0.00	0.00	0.00	0.00	0.00	0.00	0.00	0.00	1.00	0.00	0.00	0.00	1.79	1.79	6.56	0.70	0.00	0.00	0.00
[EOC₃mmor]	0.00	0.00	0.00	0.00	0.00	0.00	0.00	0.00	1.00	0.00	0.00	0.00	0.78	2.58	6.38	0.00	0.59	0.00	0.00
[MOC₃mmor]	0.00	0.00	0.00	0.00	0.00	0.00	0.00	0.00	1.00	0.00	0.00	0.00	1.42	3.12	6.47	1.38	0.00	0.00	0.00
[C₄mmor]	0.00	0.00	0.00	0.00	0.00	0.00	0.00	0.00	1.00	0.00	0.00	0.00	1.75	2.61	6.56	0.00	0.00	0.00	0.00
[EOC₁mmor]	0.00	0.00	0.00	0.00	0.00	0.00	0.00	0.00	1.00	0.00	0.00	0.00	1.85	1.94	6.71	0.49	0.00	0.00	0.00

(continued)

Table 5.7 (continued)

Abbreviation of cations/anions	A	B	C	D	E	F	G	H	I	J	K	L	M	N	O	P	Q	R	S	T	U	V	W	X	Y	Z
[C$_2$mmor]	0.00	0.00	0.00	0.00	0.00	0.00	1.00	0.00	0.00	0.00	0.00	0.00	0.00	0.00	0.00	0.00	0.00	0.00	1.63	0.86	6.56	0.00	0.00	0.00	0.00	0.00
[N$_{1,1}$CNC$_1$]	0.00	0.00	0.00	0.00	1.06	0.00	0.00	0.00	0.00	0.00	0.00	0.00	0.00	0.00	0.00	0.00	0.00	0.00	0.90	0.00	0.00	0.00	0.00	0.00	0.00	0.80
[N$_{1,1,14}$Bn]	0.00	0.00	0.00	0.00	0.96	0.00	0.00	0.00	0.00	0.00	0.00	0.00	0.00	0.00	0.00	0.00	0.00	0.00	1.05	11.7	5.70	0.00	0.00	0.88	0.00	0.00
[N$_{1,1,2}$MOC$_3$]	0.00	0.00	0.00	0.00	0.98	0.00	0.00	0.00	0.00	0.00	0.00	0.00	0.00	0.00	0.00	0.00	0.00	0.00	1.96	1.90	0.00	1.17	0.00	0.00	0.00	0.00
[C$_4$mmpy]	0.00	0.00	0.94	0.00	0.00	0.00	0.00	0.00	0.00	0.00	0.00	0.00	0.00	0.00	0.00	0.00	0.00	0.00	3.16	3.05	2.80	0.00	0.00	0.00	0.00	0.00
[C$_8$py]	0.00	0.00	1.03	0.00	0.00	0.00	0.00	0.00	0.00	0.00	0.00	0.00	0.00	0.00	0.00	0.00	0.00	0.00	1.03	1.06	5.74	0.00	0.00	0.00	0.00	0.00
[NmmC$_6$py]	0.00	0.00	1.01	0.00	0.00	0.00	0.00	0.00	0.00	0.00	0.00	0.00	0.00	0.00	0.00	0.00	0.00	0.00	2.98	5.03	3.83	0.00	0.00	0.00	0.23	0.00
[NmmC$_8$py]	0.00	0.00	1.02	0.00	0.00	0.00	0.00	0.00	0.00	0.00	0.00	0.00	0.00	0.00	0.00	0.00	0.00	0.00	2.98	1.07	3.84	0.00	0.00	0.00	0.23	0.00
[N$_{1,1,2}$MOC$_2$]	0.00	0.00	0.00	0.00	1.01	0.00	0.00	0.00	0.00	0.00	0.00	0.00	0.00	0.00	0.00	0.00	0.00	0.00	1.95	0.88	0.00	0.74	0.00	0.00	0.00	0.00
[C$_6$C$_6$pyrr]	0.00	0.00	0.00	0.00	0.00	0.00	0.00	0.00	0.00	0.00	0.00	0.00	0.00	0.00	0.00	0.00	0.00	0.00	2.08	8.89	6.59	0.00	0.00	0.00	0.00	0.00
[MOC$_2$py]	0.00	0.00	1.03	0.00	0.00	0.00	0.00	0.00	0.00	0.00	0.00	0.00	0.00	0.00	0.00	0.00	0.00	0.00	1.08	2.16	5.70	0.99	0.00	0.00	0.00	0.00
[C$_3$OHmpip]	0.00	0.00	0.00	0.00	0.00	0.00	0.00	0.00	0.00	0.00	0.00	0.00	0.00	0.00	0.00	0.00	0.00	0.00	0.72	2.80	8.04	0.00	1.02	0.00	0.00	0.00
[MOC$_3$mpyrr]	0.00	0.00	0.00	0.00	0.00	0.00	0.00	0.00	0.00	0.00	0.00	0.00	0.00	0.00	0.00	0.00	0.00	0.00	1.83	2.73	6.79	1.17	0.00	0.00	0.00	0.00
[C$_6$qui]	0.00	0.00	0.00	0.00	0.00	1.00	0.00	0.00	0.00	0.00	0.00	0.00	0.00	0.00	0.00	0.00	0.00	0.00	1.04	4.88	7.82	0.00	0.00	0.00	0.00	0.00
[MOC$_3$mim]	1.00	0.00	0.00	0.00	0.00	0.00	0.00	0.00	0.00	0.00	0.00	0.00	0.00	0.00	0.00	0.00	0.00	0.00	2.15	3.22	3.47	1.21	0.00	0.00	0.00	0.00
[CNC$_1$mim]	1.08	0.00	0.00	0.00	0.00	0.00	0.00	0.00	0.00	0.00	0.00	0.00	0.00	0.00	0.00	0.00	0.00	0.00	1.06	1.23	3.57	0.00	0.00	0.00	0.00	0.98
[C$_2$OHmmor]	0.00	0.00	0.00	0.00	0.00	0.00	1.00	0.00	0.00	0.00	0.00	0.00	0.00	0.00	0.00	0.00	0.00	0.00	0.71	1.90	6.56	3.33	0.00	0.00	0.00	0.00
[N$_{1,1,12}$Bn]	0.00	0.00	0.00	0.00	0.96	0.00	0.00	0.00	0.00	0.00	0.00	0.00	0.00	0.00	0.00	0.00	0.00	0.00	1.04	9.70	5.70	0.00	0.00	0.88	0.00	0.00
[N$_{1,1,2}$EOC$_1$]	0.00	0.00	0.00	0.00	1.03	0.00	0.00	0.00	0.00	0.00	0.00	0.00	0.00	0.00	0.00	0.00	0.00	0.00	1.99	1.12	0.00	0.38	0.00	0.00	0.00	0.00
[N$_{1,1,2}$C$_3$OH]	0.00	0.00	0.00	0.00	0.98	0.00	0.00	0.00	0.00	0.00	0.00	0.00	0.00	0.00	0.00	0.00	0.00	0.00	0.88	2.01	0.00	0.00	1.01	0.00	0.00	0.00
[N$_{1,1,2,4}$]	0.00	0.00	0.00	0.00	0.98	0.00	0.00	0.00	0.00	0.00	0.00	0.00	0.00	0.00	0.00	0.00	0.00	0.00	1.91	1.82	0.00	0.00	0.00	0.00	0.00	0.00

Group type

Anions

[BF$_4$]	0.00	0.00	0.00	0.00	0.00	0.00	0.00	0.00	0.00	0.00	0.00	0.00	0.00	0.00	0.00	0.00	0.00	0.00	0.00	0.00	0.00	0.00	0.00	0.00	0.00	0.00
[Br]	0.00	0.00	0.00	0.00	0.00	0.00	0.00	0.00	0.00	0.00	0.00	0.00	0.00	0.00	0.00	0.00	0.00	0.00	0.00	0.00	0.00	0.00	0.00	0.00	0.00	0.00
[Cl]	0.00	0.00	0.00	0.00	0.00	0.00	0.00	0.00	1.00	0.00	0.00	0.00	0.00	0.00	0.00	0.00	0.00	0.00	0.00	0.00	0.00	0.00	0.00	0.00	0.00	0.00
[DCA]	0.00	0.00	0.00	0.00	0.00	0.00	0.00	0.00	0.00	0.00	0.00	0.00	0.00	0.00	0.00	0.00	0.00	0.00	0.00	0.00	0.00	2.03	2.31
[HSO$_4$]	0.00	0.00	0.00	0.00	0.00	0.00	0.00	0.00	0.00	0.00	1.05	0.00	0.00	0.00	0.00	0.00	0.00	0.00	0.00	0.00	1.36	0.00	0.00
[I]	0.00	0.00	0.00	0.00	0.00	0.00	0.00	0.00	0.00	1.00	0.00	0.00	0.00	0.00	0.00	0.00	0.00	0.00	0.00	0.00	0.00	0.00	0.00
[MeSO$_3$]	0.00	0.00	0.00	0.00	0.00	0.00	0.00	0.00	0.00	0.00	1.02	0.00	1.15	0.00	0.00	2.16	0.00	0.00	0.00	0.00	0.00	0.00	0.00
[MeSO$_4$]	0.00	0.00	0.00	0.00	0.00	0.00	0.00	0.00	0.00	0.00	0.99	0.00	1.11	0.00	0.00	0.00	0.00	0.00	0.00	0.00	0.00	0.00	0.00
[PF$_6$]	0.00	0.00	0.00	0.00	0.00	1.00	0.00	0.00	0.00	0.00	0.00	0.00	0.00	0.00	0.00	0.00	0.00	0.00	0.00	0.00	0.00	0.00	0.00
[Tf$_2$N]	0.00	0.00	0.00	0.00	0.00	0.00	0.00	2.00	0.00	1.99	0.00	0.00	0.00	0.00	0.00	0.00	0.00	0.00	0.00	1.04	0.00	0.00	0.00
[SCN]	0.00	0.00	0.00	0.00	0.00	0.00	1.00	0.00	0.00	0.00	0.00	0.00	0.00	0.00	4.71	0.00	0.00	0.00	0.00	0.00	0.00	0.00	0.00
[Tos]	0.00	0.00	0.00	0.00	0.00	0.00	0.00	0.97	0.00	0.00	0.00	1.10	0.00	0.00	0.00	0.00	0.00	1.09	0.00	0.00	0.00	0.00	0.00
[TFA]	0.00	0.00	0.00	0.00	0.00	0.00	0.00	0.00	1.00	1.08	0.00	0.00	0.00	0.00	0.00	0.00	0.00	0.00	0.00	0.00	0.00	0.00	0.00
[TfO]	0.00	0.00	0.00	0.00	0.00	0.00	0.00	1.00	0.00	1.04	0.00	0.00	0.00	0.00	0.00	0.00	0.00	0.00	0.00	0.00	0.00	0.00	0.00
[TPTP]	0.00	0.00	0.00	1.00	0.00	0.00	0.00	0.00	0.00	0.00	0.00	0.00	0.00	0.00	0.00	0.00	0.00	0.00	0.00	0.00	0.00	0.00	0.00

Example 5.3 Use eq. **(5.5*)** to calculate $-\log EC_{50}$ for 3-(3-methoxy propyl)-1-methyl-imidazolium chloride through the traditional GC method.

Answer: The IL 3-(3-methoxypropyl)-1-methyl-imidazolium chloride has the following structure:

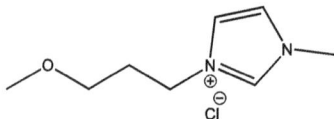

The chemical formula of the cation is $C_8H_{15}N_2O^+$. The coefficients and descriptors of this IL are given in Tables 5.5 and 5.6, respectively, which correspond to group types 1A, 1K, 2S, 3T, 3U, and 1V. The use of eq. **(5.5*)** for these parameters gives

$$-\log EC_{50}(\text{IL}, \mu M) = \beta_0 + \sum_{m=1}^{26} \beta_m X_m$$

$$= \beta_0 + \beta_1 A + \beta_{11} K + \beta_{19}(2S) + \beta_{20}(3T) + \beta_{21}(3U) + \beta_{22} V$$

$$= -5.374 - 0.084 + 1.569 - 0.048(2) + 0.321(3) - 0.119(3) - 0.057 = -3.44$$

The calculated result is close to the experimental value, that is, -3.34 [358].

5.5 Summary

Equation **(5.1*)** can assess the toxicity ($-\log EC_{50}$) of ILs toward *V. fischeri* using n_C, n_H, n_N, and n_O in cation, and n_H and n_F in the anion; two correction functions $EC_{50,IL}^+$ and $EC_{50,IL}^-$ are found from Table 5.3. Section 5.2 introduced relationships of the toxicity ($-\log EC_{50}$) of ILs toward *V. fischeri* with the structure and the 1-octanol–water partition coefficient of ILs. Equation **(5.4*)** introduced a simple GC method for the prediction of toxicity ($-\log EC_{50}$) of some ILs including imidazolium, pyridinium, and pyrrolidinium-based ILs with anions [Cl], [Br], [BF$_4$], [PF$_6$], [MeSO$_4$], [EtSO$_4$], and [DCA] toward *V. fischeri*. Equation **(5.5*)** was used for the traditional GC and NGC-2 methods to assess the toxicity ($-\log EC_{50}$) of ILs toward ICP-81. Among the different methods given in this chapter, eqs. **(5.1*)** and **(5.5*)** are recommended to assess the toxicity of ILS toward *V. fischeri* and ICP-81, respectively.

Problems

1. For [OTrop][BF$_4$], the experimental value of $-\log EC_{50}(IL, \mu M)$is -2.77 [359]. (a) Use eq. **(5.1*)** to calculate $-\log EC_{50}(IL, \mu M)$. (b) Calculate the deviation of eq. **(5.1*)** from the experimental data.

2. The experimental values of $\log K_{OW}$ and $-\log EC_{50}$ for [HMIM][PF$_6$] are -0.9508 [356] and -2.17 [357], respectively. (a) Use eq. **(5.3*)** to calculate $-\log EC_{50}(IL, \mu M)$. (b) Calculate the deviation of eq. **(5.3*)** from the experimental data.

3. Use eq. **(5.4*)** to calculate $-\log EC_{50}$ for [HMIM][Cl].

4. Use eq. **(5.5*)** to calculate $-\log EC_{50}$ for 3-(3-methoxypropyl)-1-methyl-imidazolium chloride through the NGC-2 model.

List of Symbols

A	Acceptor
AD	Applicability domain
A_{max}	Maximum acceptor superdelocalizability in a molecule
AITC	Allylisothiocyanate
ALOGP	Ghose–Crippen octanol–water partition coefficient ($A \log K_{OW,GCV}$)
ALOGP2	The squared Ghose–Crippen–Viswanadhan octanol–water partition coefficient ($A \log K_{OW,GCV}$)
Ani	Anion
ANN	Artificial neural network
AOPs	Advanced oxidation processes
ATS	Autocorrelation of a topological structure
ATS3e	Broto–Moreau autocorrelation of lag 3 (log function)/weighted by Sanderson electronegativity
ATSc2	Broto–Moreau autocorrelation descriptor of a topological structure, weighted by partial charges at lag 2 topological distance
ATSC1e	2D autocorrelation descriptor corresponding to the centered Broto–Moreau autocorrelation-lag 1/weighted by the Sanderson electronegativity
ATSC3i	Centered Broto–Moreau autocorrelation of lag 3 weighted by ionization potential
AV_{EP}	Average values of electrostatic potential
B01[C–O]	Presence/absence of C–O at topological distance 1
B02[C–C]	Presence/absence of C–C at topological distance 2
B03[N–Cl]	Presence/absence of N–Cl at topological distance 3
B04[N–O]	Presence/absence of N–O at topological distance 4
B05[C–C]	Presence/absence of C–C at topological distance 5
B06[C–C]	Presence/absence of C–C at topological distance 6
B06[C–O]	Presence/absence of C–O at topological distance 6
B07[C–N]	Presence/absence of C–N at topological distance 7
B08[C–C]	Presence/absence of C–C at topological distance 8
B08[Cl–Cl]	Descriptor with more electronegative element contents
B10[C–N]	Presence of carbon and nitrogen at the topological distance 10 (2D atom pairs)
BaA	Benzo[a]anthracene
BaP	Benzo[a]pyrene
BLTA96	Verhaar algae baseline toxicity from MLOGP (mmol L^{-1}) or log $K_{OW,M}$
BLTF96	Verhaar fish baseline toxicity from MLOGP (mmol L^{-1})
BLTD48	Verhaar Daphnia baseline toxicity from Moriguchi octanol–water partition coefficient (log $K_{OW,M}$)
BMD	Benchmark dose
BPANN	Backpropagation artificial neural network
C-015	=CH2 (atom-centered fragment)
C-026	Number of atom-centered fragments R–CX–R
C-034	Descriptor representing hydrophobic moieties
CARDD	Computer-aided rational drug design
Cat	Cation
CATS2D_03_LL	CATS2D lipophilic–lipophilic at lag 03

https://doi.org/10.1515/9783111189673-006

$CATS2D\_04\_LL$	Dragon descriptor
CCC	Concordance correlation coefficient
CD_1	Ratio of the hydrophobic volume over the total molecular surface
C_{de}	Existence of specific molecular moieties for decreasing toxicities on the basis of n_N, n_O, and n_{Hal} given in eq. (**3.11***)
C_{Hal}	Contribution of halogen atoms given in eq. (**3.6***)
C_{in}	Existence of specific molecular moieties for increasing toxicity on the basis of n_N, n_O, and n_{Hal} given in eq. (**3.11***)
$C_{Intra\ H-bond}$	Intramolecular hydrogen bonding to a hydroxyl group
C_{Iso}	Contribution of specific isomers under certain conditions given in eq. (**3.6***)
CN	1-Chloroacetophenone
CoMFA	Comparative molecular field analysis
CoMSIA	Comparative molecular similarity analysis
CR	Dibenz[b,f]-[1,4]oxazepine
CS	2-Chlorobenzylidenemalononitrile
CV	Cross-validation
$CW(S_k)$	Correlation weight of the S_k corresponding to a coefficient which is combined to the value of the descriptor if the corresponding SMILES contains S_k
CYP	Cytochrome P450 enzyme group
D	Hydrogen bond donor
2D	Two-dimensional
3D	Three-dimensional
$D604$	Number of substituted sp^2 aromatic carbons
DBPs	Disinfection by-products
$D/Dtr03$	Distance/detour ring index of order 3
$D/Dtr09$	Descriptor representing hydrophobic moieties
d_{ij}	Topological distance between two considered atoms
DFT	Density-functional theory
DISP	Decreasing isomeric structural parameters given in eqs. (**4.23***), (**4.24***), (**4.25***), and (**4.26***)
DMSO	Dimethylsulfoxide
DNEL	No-effect threshold levels for human health
D_{OW}	n-Octanol/water distribution coefficient
DSM	Decreasing structural moieties given in eq. (**3.9***)
DSTP	Decreasing structural toxicity parameters given in eq. (**2.9***)
DVT	Deep vein thrombosis
E	Electrostatic
EC_{50}	Half-maximal effective concentration
$EC_{50,JL}^-$	Decreasing toxicity by structural moieties given in eq. (**5.1***)
$EC_{50,JL}^+$	Increasing toxicity by structural moieties given in eq. (**5.1***)
$IISPE_{HOMO}$	Quantum chemistry descriptor corresponding to the highest occupied molecular orbital
E_{LUMO}	Quantum chemistry descriptor corresponding to the lowest unoccupied molecular orbital
$E_{nn}(C-C)$	Maximum nuclear–nuclear repulsion energy for a C–C bond
$E_{nn}(C-H)$	Maximum nuclear–nuclear repulsion energy for a C–H bond
EFSA	European Food Safety Authority

ERA	Environmental risk assessment
ESOL	Estimated solubility (log S) for aqueous solubility using log K_{OW} (LOGPcons)
ETA	Extended topochemical atom
ETA_dBeta	Descriptor depicting unsaturation in molecules
ETA_BetaP_S	Measurement of electronegative atom count of the molecule relative to molecular size
ETA_Epsilon_3	Descriptor showing the presence of more electronegative elements
ETA_Psi_1	A measure of hydrogen bonding propensity of the molecules and/or polar surface area
EU	European Union
F01[C–X]	Frequency of carbon and metal element at topological distance 1
F02[C–N]	Frequency of C–N at topological distance 2
F02[N–N]	Descriptor showing the presence of more electronegative elements
F03[C–N]	Frequency of C–N at topological distance 3
F02[C–S]	Frequency of C–S at topological distance 2
F03[C–S]	Frequency of C–S at topological distance 3
F03[O–P]	Frequency of O–P at topological distance 3
F04[C–N]	Frequency of C–N at topological distance 4
F04[O–O]	Frequency of O–O at topological distance 4
F05[C–O]	Frequency of C–O at topological distance 5
F07[C–C]	Frequency of C–C at topological distance 7
F07[O–O]	Frequency of O–O at topological distance 7
F09[C–S]	Frequency of C–S at topological distance 9
F09[N–N]	Descriptor with more electronegative element contents
$f_k(x)$	kth total (atom and atom-type) linear indices
$f_k^H(x)$	kth total (atom and atom-type) linear indices without considering H atoms
$f_{KL}(x_E)$	kth atom-based nonstochastic local (atom-type = heteroatoms: S, N, O) linear indices
$f_{KL}^H(x_E)$	kth atom-based nonstochastic local (atom-type = heteroatoms: S, N, O) linear indices without considering H atoms
$f_{KL}^H(x_{E-H})$	kth atom-based nonstochastic local (atom-type= H atoms bonding to heteroatoms: S, N, O) linear indices considering H atoms in the molecular pseudograph (G1)
Fr3(att)/D_D_E/1 _3s, 2_3s/	Differentiated by attraction
Fr3(rf)/A_B_B/1_2s, 2_3d/	Differentiated by refraction
Fr3(type)/C. 3_C. 3_O. 3/1_3s, 2_3s/	Differentiated by atom type
Fr3(rep)/B_D_E/1_3s 2_3d/	Differentiated by repulsion
Fr5(d_a)/A_A_A _I_I/1_4s,2_5s,3 _5d,4_5s/	Differentiated by donor–acceptor groups
Fr5(elm)/C_C_C _H_O/1_2s,2_3a, 3_5s,4_5s/	Differentiated by elemental properties
Fr5(en)/B_B_C_C _D/1_4s, 2_4s,3_4s, 3_5s/	Differentiated by electronegativity

Fr5(lip)/B_B_B_C_C/1_2s,2_4a,3_5a, 4_5a/	Differentiated by lipophilic properties
Fr5(lip)/A_C_C_C_D/1_2s,1_3s,1_5s, 4_5s/	Differentiated by lipophilic properties
G	Pauling scale
G1	Molecular pseudograph
G^2	Gravitation indexes for all bonded pairs of atoms
GA	Genetic algorithm
GA-MLR	Genetic algorithm-multiple linear regression
GATS	Geary autocorrelation descriptors
GATS1p	Geary autocorrelation-lag 1/weighted by polarizability
GATS6e	Geary autocorrelation of lag 6 weighted by Sanderson electronegativity
GATS7s	Geary autocorrelation of lag 7 weighted by I-state
GC	Group contribution
GETAWAY	GEometry, Topology, and Atom-Weights AssemblY
H	Hydrophobic
H%	Percentage of H atoms
H-047	H attached to $C1(sp^3)/C0(sp^2)$
H-051	H attached to alpha-C
HAT	Human African trypanosomiasis
HATS0e	Leverage-weighted autocorrelation of lag 0/weighted by Sanderson electronegativity
HATS1v	Leverage-weighted autocorrelation of lag 1/weighted by van der Waals volume
H bond	Hydrogen bonding functional groups given in eq. (**1.5***)
HQSAR	Hologram QSAR
Hpnotic	Dragon descriptor
Hydr	Hydrocarbon derivatives given in eq. (**1.5***)
i	Ionization potential
IARC	International Agency for Research on Cancer
I_C	A geometrical descriptor that relates to the atomic masses and the distance of the atomic nucleus from the main rotational axes
IC_{50}	Concentration producing 50% inhibition
IC_{20}	Concentration producing 20% inhibition
$(IGC_{50})^-$	Negative contribution of some specific molecular fragments and isomers given in eq. (**4.22***)
$(IC_{50})^+$	Positive contribution of some specific molecular fragments and isomers given in eq. (**4.22***)
ICP-81	Leukemia rat cell line
Ineffective	Dragon descriptor
In silico tools	Computational models that investigate pharmacological hypotheses
In vivo	Latin for "within the living"
IL	Ionic liquid
IISP	Increasing isomeric structural parameters given in eqs. (**4.23***), (**4.24***), (**4.25***), and (**4.26***)

Ip	Indicator parameter, where its value is 1 for the presence of 2- and/or 4-hydroxylated aldehyde, otherwise zero
ISM	Increasing structural moieties given in eq. (**3.9***)
$ISTP$	Increasing structural toxicity parameters given in eq. (**2.9***)
$^0\!A$	Zero-order connectivity index
$^1\!A$	First-order connectivity index
^{B}	Cross-factor ($^0\!A \times ^1\!A$)
K	Mulliken atomic electronegativity
k_H	Henry's law constant value
K_i	Inhibition constant
K_{OW}	n-Octanol/water partition coefficient
LC_{50}	Lethal concentration 50%
LD_{50}	Lethal dose concentration (50%)
$LFEA$	Global hydrogen bonding acidity of the solute
$LFEBH$	Global hydrogen bonding basicity of the solute
LMO	Leave-many-out
LOC	Lopping centric index
$LOEC$	Lowest observed effective concentration
log D	Logarithm of the ionization-corrected octanol/water partition coefficient
log $K_{OW,wc}$	Logarithm of Wildman–Crippen octanol–water partition coefficient (log K_{OW})
log P	Solvational characteristic (hydrophobicity of chemicals)
LOO	Leave-one-out
LSD	Lysergic acid diethylamide
M	Atomic mass
MAE	Mean absolute error
$MASEH$	Maximum atomic state energy for an H atom
MATS	Moran autocorrelation descriptor
$MATS6m$	Moran autocorrelation of lag 6 weighted by mass
$MATS7s$	Moran autocorrelation of lag 7 weighted by I-state
$maxdssC$	Maximum atom-type E-state: =C<
$MCS48$	Number of OQ(O)O groups in which Q is a heteroatom different from carbon or hydrogen
MD	Molecular dynamics
m_i	Atomic weights of atom i
MIM	Molecular influence matrix
$minddC$	Dragon descriptor (minimum ddC)
$mindsN$	Dragon descriptor (minimum dsN)
$minssO$	Minimum atom-type E-state: –O–
MLI	Molecular connectivity index
$MLIP$	The combinatorial descriptor that reflects molecular lipophilicity, as well as molecular size and shape
$MLOGP$ or log $K_{OW,M}$	Moriguchi octanol–water partition coefficient
$MLOGP2$	The squared Moriguchi octanol–water partition coefficient or $(\log K_{OW,M})^2$
MLR	Multiple linear regression
MO	Mustard oil
$Mor06m$	3D-MoRSE-signal 06/weighted by atomic masses

$Mor18s$	3D-MoRSE-signal 18/weighted by I-state
$Mor22s$	3D-MoRSE-signal 22/weighted by I-state
$Mor23e$	3D-MoRSE-signal 23/weighted by Sanderson electronegativity
$Mor26m$	3D-MoRSE-signal 26/weighted by atomic masses
Mp	Mean atomic polarizability (scaled on carbon atom)
$MREHO$	Maximum resonance energy for an H–O bond
MSDS	Material Safety Data Sheet
Mv	Mean atomic van der Waals volume (scaled on carbon atom)
MW	Molecular weight
Mw_{unit}	Molecular weight of repeating unit structure in g cm^{-3}
$n_{Acyclic-COO}$	Number of the acyclic ester group
NAMs	New approach methodologies
$nBondsS2$	Total number of single bonds
$N\%$	Percent of nitrogen atoms present in a molecule
$N-069$	Presence of $Ar-NH_2/X-NH_2$, where X can be O, N, S, P, Se, and halogens
n_{ArCHO}	Number of aldehydes (aromatic)
n_{ArNO2}	Number of nitro groups (aromatic)
N_b	Number of bonds in the molecule
NBs	Nitrobenzenes
n_C	Number of moles of carbon atoms
$nCIR$	Descriptor designating the presence of various rings
n_{Crs}	Number of ring secondary $C(sp^3)$
$n_{cylopentenone}$	Number of cyclopentenone ring
n_{DB}	Number of double bonds
n_{Epoxy}	Number of epoxy groups
n_F	Number of moles of fluorine atoms
NGC	Noninteger group contribution
n_H	Number of moles of hydrogen atoms
n_{Hal}	Number of moles of halogen atoms
NMLR	Nonlinear multiple regression analysis
n_N	Number of moles of nitrogen atoms
n_{NHR+NR_2}	Sum of primary and secondary amine groups
n_{NO_2}	Number of nitro groups
n_O	Number of moles of oxygen atoms
NOAEL or NOAEC	No observed adverse effect level
NOEC	No observed effective concentration
n_{OHt}	Number of tertiary alcohols
$n_{OR+1.5CN+CO}$	Sum of ether, cyano (multiplied by 1.5), and carbonyl groups
n_P	Number of moles of phosphorus atoms
$n_{Pyridines}$	Descriptor showing the presence of more electronegative elements
n_{R06}	Number of six-membered rings
n_{RCH}	Number of aldehydes (aliphatic)
n_{RNSC}	Number of isothiocyanates
n_{ROH}	Number of hydroxyl groups
NRS	Descriptor designating the presence of various rings
n_S	Number of moles of sulfur atoms
n_{Si}	Number of moles of silicon atoms

n_{sk}	Number of atoms
$n_{\text{Vinylic H except cylo}-CH_2CCOO}$	Number of vinylic hydrogen atoms except two hydrogen atoms of vinylic carbon atom (CH_2=) in the alpha position of carbonyl group attached to a cyclic ring containing ester functional group
N_{ssS}	Number of atoms of type ssS
$n_{a-CH_2-cyclo-COO}$	Number of alpha CH_2= group attached to a cyclic ring containing ester functional group
$O\%$	Percentage of O atoms
O-056	Alcohol
O-057	Phenol/enol/carboxyl OH
O-058	=O group (sp^2 hybridization) in a molecule
O-060	Al-O-Ar/Ar-O-Ar/R. . .O. . .R/R-O-C=X
P	Atomic polarizability
PAHs	Polycyclic aromatic hydrocarbons
PCA	Principal component analysis
PE	Pulmonary embolism
PE1	Polarization energy x electronic energy
$pEC^-_{50,CR}$	Structural descriptor corresponding to various substituents for decreasing the value of –log EC_{50} of eq. (**2.21***) in different positions of 11*H*-dibenz[*b,e*]azepine and dibenz[*b,f*][1,4]-oxazepine derivatives
$pEC^+_{50,CR}$	Structural descriptor corresponding to various substituents for increasing the value of –log EC_{50} of eq. (**2.21***) in different positions of 11*H*-dibenz[*b,e*]azepine and dibenz[*b,f*][1,4]-oxazepine derivatives
PG	Polar groups given in eq. (**1.5***)
PHB	Polyhydroxybenzene
pIC^-_{50} and PIC^+_{50}	Nonadditive influences of some specific groups in certain positions of the framework of different classes given in eq. (**3.12***)
PNEC	The concentration of a chemical that marks the limit at which below no adverse effects of exposure in an ecosystem are measured
PLS	Partial least-squares method
PNSA_2/TMSA	FNSA_2 fractional PNSA, which calculates atomic partial charges to the total molecular solvent-accessible surface area
P_O	Valency-related descriptor, which relates to the strength of intermolecular bonding interactions
P_{SIGMA}	Maximum bond order for a given pair of atomic species in the molecule
Psi_i_1d	Intrinsic state pseudoconnectivity index type 1d
P_VSA_i_1	P_VSA-like on ionization potential, bin 1
PW5	Path/walk 5 – Randic shape index
q^2	A mean cross-validated r^2
q^2_{F1}, q^2_{F2}, q^2_{F3}	Extra external validation coefficients
q^2_{LOO}	r^2 of leave-one-out
q^2_{LMO}	r^2 of leave-many-out
QSAR	Quantitative structure–activity relationship
QSPR	Quantitative structure–property relationship
QSTR	Quantitative structure–toxicity relationship
QSTTR	Quantitative structure–toxicity–toxicity relationship
Q_x	Total charge of halogens in a molecule
3R rules	Reduction, replacement, and refinement

R	Rugosity descriptor
r^2	Coefficient of determination
$R1e$	R autocorrelation of lag 1/weighted by Sanderson electronegativity
$R3u+$	R maximal autocorrelation of lag 3/unweighted
Rana japonica	Japanese brown frog
R_{C-C}	Distance between two carbon atoms
$RDCHI$	Descriptor depicting unsaturation in molecules
REACH	Registration, Evaluation, Authorization and Restriction of Chemicals
RfD	Reference dose
r_{ij}	Interatomic distance
RMSE	Root mean square error
RNH	Relative number of H atoms
RPCG	Relative positive charge, which belongs to electrostatic descriptors
s	E-state; electrotopological states
S	Steric
S3K	Path Kier alpha-modified shape index
SaaaC	Descriptor representing hydrophobic moieties
S_A(att)/E_E_E_F/1_4s,2_3a/3	Differentiated by attraction properties
S_A(chg)/A_C_C_D/1_2s, 1_4s,3_4s/6	Differentiated by charge
S_A(lip)/B_B_C_C/2_4s,3_4s/	Differentiated by lipophilic properties
S_A(rep)/B_C_C_C/1_3s,1_4s/4	Differentiated by repulsion properties
S_A(type)/C.3_C.3_H_O.3/1_2 s, 2_4 s,3_4 s/6	Differentiated by atom type
SAacc	Surface area of acceptor atoms from P_VSA-like descriptors
SAG	Surface area grid
SAR	Structure–activity relationship
S_A(att)/E_E_E_F/ 1_4s,2_3a/3	Differentiated by attraction properties
S_A(type)/C.1_C.1_C.3_H/2_3s, 3_4s/4	Differentiated by elemental properties
S-BRC	Single-benzene ring compounds
S_{EP}	Electrostatic potential surface area
S_k	A SMILES atom, that is, one symbol (e.g., "C," "c," "N," and "O") or a group of symbols that cannot be examined separately (e.g., "Cl" and "Br")
SKIN	Skin permeability descriptor
SPH	Spherosity
$SssCH_2$	Sum of $ssCH_2$ E-states
SsssCH	Descriptor influencing branching in a molecule
STLs	Sesquiterpene lactones
Sub	Substitution
SubC1	Number of primary carbons
SVM	Support vector machine
T_{25}	The chronic daily dose in mg per kg bodyweight, which will give 25% of the animals' tumors at a specific tissue site
TA100	*Salmonella typhimurium* TA100 strain
TA98+S9	Mutagenicity against *Salmonella typhimurium* (TA98) bacterial species with implementing microsomal-activating enzyme named S9

TA98-S9	Mutagenicity against *Salmonella typhimurium* (TA98) bacterial species without implementing microsomal-activating enzyme named S9
Tbr	*Trypanosoma brucei rhodesiense*
$T(Br. . .Br)$	Sum of topological distances between Br. . .Br
$T(N. . .F)$	Sum of topological distances between N. . .F
$T(O. . .O)$	Sum of topological distances between two oxygen atoms
TE or E_T	Total energy
TI	Thrombin inhibitor
TI^+ and TI^-	Nonadditive influences of some groups and molecular moieties given in eq. (**3.13***)
TIC2	Dragon descriptor
ToPSA	Topological polar surface area
TOMOCOMD	TOpological MOlecular COMputer Design
Tox^-	Contribution of some molecular moieties can decrease the toxicity of a nitroaromatic compound given in eq. (**2.4***)
Tox^+	Contribution of some molecular moieties can increase the toxicity of a nitroaromatic compound given in eq. (**2.4***)
TPSA(NO)	Topological polar surface area using N, O polar contributions
TPSA(Tot)	Topological polar surface area using N, O, S, P polar contributions
TRP	Transient receptor potential
TRPA1	Transient receptor potential ankyrin 1
TU_0	Toxicity units
$TU_{1/2}$	Half-life toxicity
Ui	Unsaturation index
UPS	UV photoelectron spectroscopy
US EPA	United States Environmental Protection Agency
V or *v*	van der Waals atomic volume
V/A/P	Vinyl/allyl/propargyl
VOCs	Volatile organic compounds
$\overline{W}$	Average value of atomic polarizability
W_3	Hydrophilic region descriptor
WHIM descriptors	Weighted holistic invariant molecular descriptors
WN_1 and WN_5	H-bond acceptor volume descriptors
WWTPs	Wastewater treatment plants
X0sol	Solvation connectivity index of order 0
X1sol	Solvation connectivity index of order 1
X1A	Average connectivity index of order 1
X3A	Average connectivity index of order 3
X3Av	Average valence connectivity index of order 3
X5Av	Descriptor with more hydrophobic influences
X5sol	Solvation connectivity index of order 5
YS	Y-scrambling
Z_C and Z_H	Nuclear (core) charges of atoms C and H, respectively
γ	Surface tension
δ_{ij}	Kronecker delta
δ	Hildebrand solubility parameter
δ_D	Dispersion component of Hildebrand parameter δ
δ_H	Hydrogen bonding component of Hildebrand parameter δ

δ_P	Polar component of Hildebrand parameter δ
δ^{Inc} and δ^{Dec}	Two correcting functions given in eq. (1.3*) for adjustment of the underestimating and overestimating results from the contribution of elemental composition as compared to experimental data
$x1A$	The selected connectivity descriptor
μ	Dipole moment
$\Delta9$-THC	$\Delta9$-Tetrahydrocannabinol
$\Delta_f H(g)$	Gas-phase heat of formation
ϖ	Electrophilicity index
η	Hardness

Answers to Problems

Chapter 1

1. (a) 42.92 MPa$^{1/2}$
   (b) 25.57 MPa$^{1/2}$
   (c) 42.92 MPa$^{1/2}$
   (d) 28.31 MPa$^{1/2}$
2. (a) −5.16
   (b) −3.08
   (c) −4.67
   (d) −7.26

Chapter 2

1. (a) −log (LD_{50}) = −4.695
   (b) Equation (**2.4***)
2. (a) 0.374
   (b) −0.369
3. Equation (**2.7***): −log $IGC_{50}(/mM)$ for 3-ethoxy-4-hydroxybenzaldehyde, and 3-hydroxy-4-methoxybenzaldehyde are 0.336 (Dev = 0.321) and −0.218 (Dev = −0.076), respectively. Deviations of the predicted −log $IGC_{50}(/mM)$ for 3-ethoxy-4-hydroxybenzaldehyde and 3-hydroxy-4-methoxybenzaldehyde from Problem 2 are 0.359 and −0.227, respectively. Thus, eq. (**2.7***) gives closer prediction for both compounds as compared to the experimental data.
4. 0.708; Dev = 0.091
5. −3.295
6. (a) Equation (**2.12***): −0.43; eq. (**2.13***): −0.31
   (b) Equation (**2.13***)
7. (a) Equation (**2.14***): 3.14; eq. (**2.15***): 3.28; eq. (**2.16***): 3.27.
   (b) Equation (**2.14***): Dev = −0.44; eq. (**2.15***): Dev = −0.30; eq. (**2.16***): Dev = −0.31. Thus, eqs. (**2.15***) and (**2.16***) give more reliable results as compared to eq. (**2.14***).
8. (a) 7.006
   (b) Equation (**2.17***): Dev = −0.006

CoMFA: Dev = 0.018
CoMSIA: Dev = −0.028

https://doi.org/10.1515/9783111189673-007

Chapter 3

1. (a) 0.33
   (b) −0.09
2. (a) −6.976
   (b) Deviation of eq. (**3.8***)=2.970
   Deviation of eq. (**3.9***)=−0.716
   Thus, the predicted result of eq. (**3.9***) is very close to the experimental data.
3. (a) −6.01
   (b) Deviation of eq. (**3.10***)=−1.28
   Deviation of eq. (**3.11***)=0.2
   Thus, the predicted result of eq. (**3.11***) is very close to the experimental data.
4. 4.388 where $pIC_{50}^{-} = 1.0$
5. (a) 3.05
   (b) Equation (**3.13***): Dev = −0.06
   Mena-Ulecia et al.: Dev = 0.24
6. (a) Equation (**3.16***): 2.176
   Equation (**3.14***): 2.70
   Equation (**3.15***): 2.69
   (b) Equation (**3.16***) gives a closer prediction.

Chapter 4

1. (a) 2.2410, Oc2ccc(Oc1ccccc1)cc2
   (b) 3.5160
   (c) 0.5140
2. (a) 2.14
   (b) 0.52
   (c) Su et al.
3. The values of $-\log IC_{50}$, $-\log IC_{20}$, $-\log LOEC$, and $-\log NOEC$ toward *C. vulgaris* are 1.284, 1.355, 1.400, and 1.692, respectively.

Chapter 5

1. (a) −2.05
    (b) 0.72
2. (a) −2.36
    (b) −0.19
3. −2.15
4. −3.36

References

[1]   D. SE, T.J. Schöpf, N. Renner, Regulation (EC) No. 1907/2006 of the European Parliament and of the
       Council of 18 December 2006 concerning the Registration, Evaluation, Authorisation and Restriction
       of Chemicals (REACH), http://eur-lex.europa.eu/legal-content/EN/TXT/PDF/?uri=CELEX:02006R1907-
       20161011&from=EN (2019).
[2]   E. Walum, Acute oral toxicity, Environmental Health Perspectives, 106 (1998) 497–503.
[3]   J. Strickland, A.J. Clippinger, J. Brown, D. Allen, A. Jacobs, J. Matheson, A. Lowit, E.N. Reinke,
       M.S. Johnson, M.J. Quinn Jr, Status of acute systemic toxicity testing requirements and data uses by
       US regulatory agencies, Regulatory Toxicology and Pharmacology, 94 (2018) 183–196.
[4]   G.J. Myatt, E. Ahlberg, Y. Akahori, D. Allen, A. Amberg, L.T. Anger, A. Aptula, S. Auerbach, L. Beilke,
       P. Bellion, In silico toxicology protocols, Regulatory Toxicology and Pharmacology, 96 (2018) 1–17.
[5]   M. Sigurnjak Bureš, M. Cvetnić, M. Miloloža, D. Kučić Grgić, M. Markić, H. Kušić, T. Bolanča,
       M. Rogošić, Š. Ukić, Modeling the toxicity of pollutants mixtures for risk assessment: a review,
       Environmental Chemistry Letters, 19 (2021) 1629–1655.
[6]   X.H. Wang, L.Y. Fan, S. Wang, Y. Wang, L.C. Yan, S.S. Zheng, C.J. Martyniuk, Y.H. Zhao, Relationship
       between acute and chronic toxicity for prevalent organic pollutants in Vibrio fischeri based upon
       chemical mode of action, Journal of Hazardous Materials, 338 (2017) 458–465.
[7]   H.J. Verhaar, C.J. Van Leeuwen, J.L. Hermens, Classifying environmental pollutants, Chemosphere, 25
       (1992) 471–491.
[8]   O. Isayev, B. Rasulev, L. Gorb, J. Leszczynski, Structure-toxicity relationships of nitroaromatic
       compounds, Molecular Diversity, 10 (2006) 233–245.
[9]   J. Verma, V.M. Khedkar, E.C. Coutinho, 3D-QSAR in drug design – a review, Current Topics in
       Medicinal Chemistry, 10 (2010) 95–115.
[10]   K. Roy, S. Kar, R.N. Das, A Primer on QSAR/QSPR Modeling: Fundamental Concepts, Springer, (2015).
[11]   P. Gramatica, N. Chirico, E. Papa, S. Cassani, S. Kovarich, QSARINS: a new software for the
       development, analysis, and validation of QSAR MLR models, Journal of Computational Chemistry, 34
       (2013) 2121–2132. http://www.qsar.it.
[12]   J. Shao, Linear model selection by cross-validation, Journal of the American Statistical Association,
       88 (1993) 486–494.
[13]   S. Geisser, The predictive sample reuse method with applications, Journal of the American Statistical
       Association, 70 (1975) 320–328.
[14]   K. Roy, I. Mitra, On various metrics used for validation of predictive QSAR models with applications
       in virtual screening and focused library design, Combinatorial Chemistry & High Throughput
       Screening, 14 (2011) 450–474.
[15]   C. Rücker, G. Rücker, M. Meringer, y-Randomization and its variants in QSPR/QSAR, Journal of
       Chemical Information and Modeling, 47 (2007) 2345–2357.
[16]   R. Veerasamy, H. Rajak, A. Jain, S. Sivadasan, C.P. Varghese, R.K. Agrawal, Validation of QSAR
       models-strategies and importance, International Journal of Drug Design & Discovery, 3 (2011)
       511–519.
[17]   P.P. Roy, J.T. Leonard, K. Roy, Exploring the impact of size of training sets for the development of
       predictive QSAR models, Chemometrics and Intelligent Laboratory Systems, 90 (2008) 31–42.
[18]   K. Roy, I. Mitra, S. Kar, P.K. Ojha, R.N. Das, H. Kabir, Comparative studies on some metrics for
       external validation of QSPR models, Journal of Chemical Information and Modeling, 52 (2012)
       396–408.
[19]   A. Golbraikh, A. Tropsha, Beware of Q2, Journal of Molecular Graphics and Modelling, 20 (2002)
       269–276.

https://doi.org/10.1515/9783111189673-008

[20] A. Golbraikh, M. Shen, Z. Xiao, Y.-D. Xiao, K.-H. Lee, A. Tropsha, Rational selection of training and test sets for the development of validated QSAR models, Journal of Computer-Aided Molecular Design, 17 (2003) 241–253.

[21] R. Garg, C.J. Smith, Predicting the bioconcentration factor of highly hydrophobic organic chemicals, Food and Chemical Toxicology, 69 (2014) 252–259.

[22] P. Gramatica, S. Cassani, N. Chirico, QSARINS chem: Insubria datasets and new QSAR/QSPR models for environmental pollutants in QSARINS, Journal of Computational Chemistry, 35 (2014) 1036–1044. http://www.qsar.it.

[23] L.M. Shi, H. Fang, W. Tong, J. Wu, R. Perkins, R.M. Blair, W.S. Branham, S.L. Dial, C.L. Moland, D.M. Sheehan, QSAR models using a large diverse set of estrogens, Journal of Chemical Information and Computer Sciences, 41 (2001) 186–195.

[24] G. Schüürmann, R.-U. Ebert, J. Chen, B. Wang, R. Kühne, External validation and prediction employing the predictive squared correlation coefficient – Test set activity mean vs training set activity mean, Journal of Chemical Information and Modeling 48 (2008) 2140–2145.

[25] V. Consonni, D. Ballabio, R. Todeschini, Comments on the definition of the Q2 parameter for QSAR validation, Journal of Chemical Information and Modeling, 49 (2009) 1669–1678.

[26] L. L.I.-K., Assay validation using the concordance correlation coefficient, Biometrics, (1992) 599–604.

[27] L. L.I.-K., A concordance correlation coefficient to evaluate reproducibility, Biometrics, (1989) 255–268.

[28] K. Roy, P. Chakraborty, I. Mitra, P.K. Ojha, S. Kar, R.N. Das, Some case studies on application of "rm2" metrics for judging quality of quantitative structure–activity relationship predictions: emphasis on scaling of response data, Journal of Computational Chemistry, 34 (2013) 1071–1082.

[29] N. Chirico, P. Gramatica, Real external predictivity of QSAR models: how to evaluate it? Comparison of different validation criteria and proposal of using the concordance correlation coefficient, Journal of Chemical Information and Modeling, 51 (2011) 2320–2335.

[30] P. Gramatica, S. Cassani, P.P. Roy, S. Kovarich, C.W. Yap, E. Papa, QSAR Modeling is not "Push a Button and Find a Correlation": a case study of toxicity of (benzo) triazoles on algae, Molecular Informatics, 31 (2012) 817–835.

[31] E. Papa, S. Kovarich, P. Gramatica, Development, validation and inspection of the applicability domain of QSPR models for physicochemical properties of polybrominated diphenyl ethers, QSAR Combinatorial Science, 28 (2009) 790–796.

[32] M.H. Kutner, C. Nachtsheim, J. Neter, Applied Linear Regression Models, McGraw-Hill/Irwin, (2004).

[33] T. Puzyn, J. Leszczynski, M.T. Cronin, Recent Advances in QSAR Studies: Methods and Applications, Springer Science & Business Media, (2010).

[34] M. Meloun, J. Militký, M. Hill, R.G. Brereton, Crucial problems in regression modelling and their solutions, Analyst, 127 (2002) 433–450.

[35] M. Meloun, S. Bordovská, K. Kupka, Outliers detection in the statistical accuracy test of a pK a prediction, Journal of Mathematical Chemistry, 47 (2010) 891–909.

[36] P. Gramatica, On the Development and Validation of QSAR Models, in: Computational Toxicology, Springer, (2013), pp. 499–526.

[37] J. Jaworska, N. Nikolova-Jeliazkova, T. Aldenberg, QSAR applicability domain estimation by projection of the training set descriptor space: a review, ATLA, 33 (2005) 445–459.

[38] D.R. Joshi, N. Adhikari, An overview on common organic solvents and their toxicity, Journal of Pharmaceutical Research International, 28 (2019) 1–18.

[39] P. Rai, S. Mehrotra, S. Priya, E. Gnansounou, S.K. Sharma, Recent advances in the sustainable design and applications of biodegradable polymers, Bioresource Technology, (2021) 124739.

[40] F. Lima, C.H. Dietz, A.J. Silvestre, L.C. Branco, J. Canongia Lopes, F. Gallucci, K. Shimizu, C. Held, I.M. Marrucho, Vapor pressure assessment of sulfolane-based eutectic solvents: experimental, PC-SAFT, and molecular dynamics, The Journal of Physical Chemistry B, 124 (2020) 10386–10397.

[41] H. Matsukawa, M. Otsuka, K. Otake, Phase-equilibrium measurement of a carbon dioxide/toluene/ polystyrene ternary system using laser turbidimetry, Fluid Phase Equilibria, 509 (2020) 112464.

[42] J. Moon, H.E. Yang, C.H. Lee, J.S. Choi, J.S. Oh, Phase equilibria and surface tension in castor oil-based polyols-water–methanol mixture: Thermodynamic basis, Journal of Applied Polymer Science, 138 (2021) 50101.

[43] M. Pirdashti, Z. Heidari, N. Abbasi Fashami, S.M. Arzideh, I. Khoiroh, Phase equilibria of aqueous two-phase systems of PEG with sulfate salt: effects of pH, temperature, type of cation, and polymer molecular weight, Journal of Chemical & Engineering Data, 66 (2021) 1425–1434.

[44] A. Samarov, L. Shishaeva, A. Toikka, Phase equilibria and extraction properties of deep eutectic solvents in alcohol–ester systems, Theoretical Foundations of Chemical Engineering, 54 (2020) 551–559.

[45] D.M. Walden, Y. Bundey, A. Jagarapu, V. Antontsev, K. Chakravarty, J. Varshney, Molecular simulation and statistical learning methods toward predicting drug–polymer amorphous solid dispersion miscibility, stability, and formulation design, Molecules, 26 (2021) 182.

[46] J. Wu, B. Wu, W. Wang, K.S. Chiang, A.K.-Y. Jen, J. Luo, Ultra-efficient and stable electro-optic dendrimers containing supramolecular homodimers of semifluorinated dipolar aromatics, Materials Chemistry Frontiers, 2 (2018) 901–909.

[47] S. Venkatram, C. Kim, A. Chandrasekaran, R. Ramprasad, Critical assessment of the hildebrand and hansen solubility parameters for polymers, Journal of Chemical Information and Modeling, 59 (2019) 4188–4194.

[48] S.D. Bergin, Z. Sun, D. Rickard, P.V. Streich, J.P. Hamilton, J.N. Coleman, Multicomponent solubility parameters for single-walled carbon nanotube– solvent mixtures, ACS Nano, 3 (2009) 2340–2350.

[49] Q.-Y. Li, Z.-F. Yao, J.-Y. Wang, J. Pei, Multi-level aggregation of conjugated small molecules and polymers: from morphology control to physical insights, Reports on Progress in Physics, 31 (2021) 84.

[50] J. Brandrup, E.H. Immergut, E.A. Grulke, A. Abe, D.R. Bloch, Polymer Handbook, Wiley, New York, (1999).

[51] J. Stipek, H. Daoust, Additives for Plastics, Springer Science & Business Media, New York, (2012).

[52] F. Rodríguez-Ropero, T. Hajari, N.F. van der Vegt, Mechanism of polymer collapse in miscible good solvents, The Journal of Physical Chemistry B, 119 (2015) 15780–15788.

[53] M. Tobiszewski, S. Tsakovski, V. Simeonov, J. Namieśnik, F. Pena-Pereira, A solvent selection guide based on chemometrics and multicriteria decision analysis, Green Chemistry, 17 (2015) 4773–4785.

[54] K. Alfonsi, J. Colberg, P.J. Dunn, T. Fevig, S. Jennings, T.A. Johnson, H.P. Kleine, C. Knight, M.A. Nagy, D.A. Perry, Green chemistry tools to influence a medicinal chemistry and research chemistry based organisation, Green Chemistry, 10 (2008) 31–36.

[55] X. Yu, X. Wang, H. Wang, X. Li, J. Gao, Prediction of solubility parameters for polymers by a QSPR model, QSAR & Combinatorial Science, 25 (2006) 156–161.

[56] D.W. Van Krevelen, K. Te Nijenhuis, Properties of Polymers: Their Correlation with Chemical Structure; Their Numerical Estimation and Prediction from Additive Group Contributions, Elsevier, Amsterdam, 2009.

[57] N. Goudarzi, M.A. Chamjangali, A. Amin, Calculation of Hildebrand solubility parameters of some polymers using QSPR methods based on LS-SVM technique and theoretical molecular descriptors, Chinese Journal of Polymer Science, 32 (2014) 587–594.

[58] D.İ. Koç, M.L. Koç, A genetic programming-based QSPR model for predicting solubility parameters of polymers, Chemometrics and Intelligent Laboratory Systems, 144 (2015) 122–127.

[59] D.İ. Koç, M.L. Koç, QSPR prediction of polymers' solubility parameters by radial basis functional link net, Journal of Computational Methods in Sciences and Engineering, (2020) 1–16.

[60] M.H. Keshavarz, M. Shafiee, B.N. Jazi, Simple Approach for Reliable Prediction of Solubility of Polymers in Environmentally Compatible Solvents, Industrial & Engineering Chemistry Research, 61 (2022) 2425–2433.

[61] T.M. Klapötke, Chemistry of High-Energy Materials, 6th ed., Walter de Gruyter GmbH & Co KG, 2022.

[62] P. Marchetti, M.F. Jimenez Solomon, G. Szekely, A.G. Livingston, Molecular separation with organic solvent nanofiltration: a critical review, Chemical Reviews, 114 (2014) 10735–10806.

[63] C.D. Klaassen, Casarett and Doull's toxicology: The Basic Science of Poisons, McGraw-Hill, New York (2013).

[64] M. Honda, N. Suzuki, Toxicities of polycyclic aromatic hydrocarbons for aquatic animals, International Journal of Environmental Research and Public Health, 17 (2020) 1363.

[65] K. Hayakawa, Oil spills and polycyclic aromatic hydrocarbons, in: K. Hayakawa (Ed.) Polycyclic Aromatic Hydrocarbons, Springer, Singapore, (2018), pp. 213–223.

[66] I.C. Romero, T. Sutton, B. Carr, E. Quintana-Rizzo, S.W. Ross, D.J. Hollander, J.J. Torres, Decadal assessment of polycyclic aromatic hydrocarbons in mesopelagic fishes from the Gulf of Mexico reveals exposure to oil-derived sources, Environmental Science & Technology, 52 (2018) 10985–10996.

[67] N.L. Devi, I.C. Yadav, Q. Shihua, Y. Dan, G. Zhang, P. Raha, Environmental carcinogenic polycyclic aromatic hydrocarbons in soil from Himalayas, India: Implications for spatial distribution, sources apportionment and risk assessment, Chemosphere, 144 (2016) 493–502.

[68] K. Bekki, A. Toriba, N. Tang, T. Kameda, K. Hayakawa, Biological effects of polycyclic aromatic hydrocarbon derivatives, Journal of UOEH, 35 (2013) 17–24.

[69] G.N. Cherr, E. Fairbairn, A. Whitehead, Impacts of petroleum-derived pollutants on fish development, The Annual Review of Animal Biosciences, 5 (2017) 185–203.

[70] S.G. Machatha, S.H. Yalkowsky, Comparison of the octanol/water partition coefficients calculated by ClogP®, ACDlogP and KowWin® to experimentally determined values, International Journal of Pharmaceutics, 294 (2005) 185–192.

[71] Z. Wang, Y. Su, W. Shen, S. Jin, J.H. Clark, J. Ren, X. Zhang, Predictive deep learning models for environmental properties: the direct calculation of octanol–water partition coefficients from molecular graphs, Green Chemistry, 21 (2019) 4555–4565.

[72] H. Cumming, C. Rücker, Octanol–water partition coefficient measurement by a simple 1H NMR method, ACS Omega, 2 (2017) 6244–6249.

[73] O.K. Dalrymple, Experimental determination of the octanol-water partition coefficient for acetophenone and atrazine, Physical & Chemical Principles of Environmental Engineering, 3 (2005) 1–7.

[74] P. Molyneux, Octanol/water partition coefficients Kow: A critical examination of the value of the methylene group contribution to log Kow for homologous series of organic compounds, Fluid Phase Equilibria, 368 (2014) 120–141.

[75] A. Daina, O. Michielin, V. Zoete, iLOGP: a simple, robust, and efficient description of n-octanol/water partition coefficient for drug design using the GB/SA approach, Journal of Chemical Information and Modeling, 54 (2014) 3284–3301.

[76] K.B. Hanson, D.J. Hoff, T.J. Lahren, D.R. Mount, A.J. Squillace, L.P. Burkhard, Estimating n-octanol-water partition coefficients for neutral highly hydrophobic chemicals using measured n-butanol-water partition coefficients, Chemosphere, 218 (2019) 616–623.

[77] V. Kundi, J. Ho, Predicting octanol–water partition coefficients: are quantum mechanical implicit solvent models better than empirical fragment-based methods? The Journal of Physical Chemistry B, 123 (2019) 6810–6822.

[78] X. Kang, B. Hu, M.C. Perdana, Y. Zhao, Z. Chen, Extreme learning machine models for predicting the n-octanol/water partition coefficient (Kow) data of organic compounds, Journal of Environmental Chemical Engineering, 10 (2022) 108552.

[79] T. Sekiguchi, K. Yachiguchi, M. Kiyomoto, S. Ogiso, S. Wada, Y. Tabuchi, C.-S. Hong, A.K. Srivastav, S.D. Archer, S.B. Pointing, Molecular mechanism of the suppression of larval skeleton by polycyclic aromatic hydrocarbons in early development of sea urchin Hemicentrotus pulcherrimus, Fisheries Science, 84 (2018) 1073–1079.

[80] L. Dsikowitzky, I. Nordhaus, N. Andarwulan, H.E. Irianto, H.N. Lioe, F. Ariyani, S. Kleinertz, J. Schwarzbauer, Accumulation patterns of lipophilic organic contaminants in surface sediments and in economic important mussel and fish species from Jakarta Bay, Indonesia, Marine Pollution Bulletin, 110 (2016) 767–777.

[81] A.R. Jafarabadi, A.R. Bakhtiari, Z. Yaghoobi, C.K. Yap, M. Maisano, T. Cappello, Distributions and compositional patterns of polycyclic aromatic hydrocarbons (PAHs) and their derivatives in three edible fishes from Kharg coral Island, Persian Gulf, Iran, Chemosphere, 215 (2019) 835–845.

[82] S. Kumar, G. Kaushik, J.F. Villarreal-Chiu, Scenario of organophosphate pollution and toxicity in India: a review, Environmental Science and Pollution Research, 23 (2016) 9480–9491.

[83] G.K. Sidhu, S. Singh, V. Kumar, D.S. Dhanjal, S. Datta, J. Singh, Toxicity, monitoring and biodegradation of organophosphate pesticides: a review, Critical Reviews in Environmental Science and Technology, 49 (2019) 1135–1187.

[84] A.K. Greaves, R.J. Letcher, A review of organophosphate esters in the environment from biological effects to distribution and fate, Bulletin of Environmental Contamination and Toxicology, 98 (2017) 2–7.

[85] V. Kumar, N. Upadhay, A. Wasit, S. Singh, P. Kaur, Spectroscopic methods for the detection of organophosphate pesticides – a preview, Current World Environment, 8 (2013) 313.

[86] A.M. King, C.K. Aaron, Organophosphate and carbamate poisoning, Emergency Medicine Clinics, 33 (2015) 133–151.

[87] L. Sun, W. Xu, T. Peng, H. Chen, L. Ren, H. Tan, D. Xiao, H. Qian, Z. Fu, Developmental exposure of zebrafish larvae to organophosphate flame retardants causes neurotoxicity, Neurotoxicology and Teratology, 55 (2016) 16–22.

[88] T. Pettit, M. Bettes, A.R. Chapman, L.M. Hoch, N.D. James, P.J. Irga, F.R. Torpy, The botanical biofiltration of VOCs with active airflow: is removal efficiency related to chemical properties? Atmospheric Environment, 214 (2019) 116839.

[89] L. Allou, L. El Maimouni, S. Le Calvé, Henry's law constant measurements for formaldehyde and benzaldehyde as a function of temperature and water composition, Atmospheric Environment, 45 (2011) 2991–2998.

[90] D. Ghernaout, Aeration process for removing radon from drinking water – a review, Applied Engineering, 3 (2019) 32–45.

[91] H. Shen, Z. Chen, H. Li, X. Qian, X. Qin, W. Shi, Gas-particle partitioning of carbonyl compounds in the ambient atmosphere, Environmental Science & Technology, 52 (2018) 10997–11006.

[92] T. Chmiel, A. Mieszkowska, D. Kempińska-Kupczyk, A. Kot-Wasik, J. Namieśnik, Z. Mazerska, The impact of lipophilicity on environmental processes, drug delivery and bioavailability of food components, Microchemical Journal, 146 (2019) 393–406.

[93] A.E. Gorji, Z.E. Gorji, S. Riahi, Quantitative structure-property relationship (QSPR) for prediction of CO_2 Henry's law constant in some physical solvents with consideration of temperature effects, Korean Journal of Chemical Engineering, 34 (2017) 1405–1415.

[94] U. EPA, Estimation Programs Interface SuiteTM for Microsoft® Windows, in, United States Environmental Protection Agency, Washington, DC, USA, (2016).

[95] X. Wang, Z. Guo, Z. Liu, X. Zhang, Curvature dependence of Henry's law constant and nonideality of gas equilibrium for curved vapor–liquid interfaces, AIChE Journal, 65 (2019) e16604.

[96] P.R. Duchowicz, J.C. Garro, E.A. Castro, QSPR study of the Henry's Law constant for hydrocarbons, Chemometrics and Intelligent Laboratory Systems, 91 (2008) 133–140.

[97]   F. Gharagheizi, R. Abbasi, B. Tirandazi, Prediction of Henry's law constant of organic compounds in water from a new group-contribution-based model, Industrial & Engineering Chemistry Research, 49 (2010) 10149–10152.

[98]   M. Goodarzi, E.V. Ortiz, L.d.S. Coelho, P.R. Duchowicz, Linear and non-linear relationships mapping the Henry's law parameters of organic pesticides, Atmospheric Environment, 44 (2010) 3179–3186.

[99]   H. Modarresi, H. Modarress, J. Dearden, Henry's law constant of hydrocarbons in air–water system: The cavity ovality effect on the non-electrostatic contribution term of solvation free energy, SAR and QSAR in Environmental Research, 16 (2005) 461–482.

[100] H. Modarresi, H. Modarress, J.C. Dearden, QSPR model of Henry's law constant for a diverse set of organic chemicals based on genetic algorithm-radial basis function network approach, Chemosphere, 66 (2007) 2067–2076.

[101] D.R. O'Loughlin, N.J. English, Prediction of Henry's Law Constants via group-specific quantitative structure property relationships, Chemosphere, 127 (2015) 1–9.

[102] N.K. Razdan, D.M. Koshy, J.M. Prausnitz, Henry's constants of persistent organic pollutants by a group-contribution method based on scaled-particle theory, Environmental Science & Technology, 51 (2017) 12466–12472.

[103] P.R. Duchowicz, J.F. Aranda, D.E. Bacelo, S.E. Fioressi, QSPR study of the Henry's law constant for heterogeneous compounds, Chemical Engineering Research and Design, 154 (2020) 115–121.

[104] M.H. Keshavarz, M. Rezaei, S.H. Hosseini, A simple approach for prediction of Henry's law constant of pesticides, solvents, aromatic hydrocarbons, and persistent pollutants without using complex computer codes and descriptors, Process Safety and Environmental Protection, 162 (2022) 867–877.

[105] T. Huang, G. Sun, L. Zhao, N. Zhang, R. Zhong, Y. Peng, Quantitative structure-activity relationship (QSAR) studies on the toxic effects of nitroaromatic compounds (NACs): A systematic review, International Journal of Molecular Sciences, 22 (2021) 8557.

[106] J. Tiwari, P. Tarale, S. Sivanesan, A. Bafana, Environmental persistence, hazard, and mitigation challenges of nitroaromatic compounds, Environmental Science and Pollution Research, 26 (2019) 28650–28667.

[107] Y. Deng, R. Zhao, Advanced oxidation processes (AOPs) in wastewater treatment, Current Pollution Reports, 1 (2015) 167–176.

[108] J. Min, J. Wang, W. Chen, X. Hu, of 2-chloro-4-nitrophenol via a hydroxyquinol pathway by a Gram-negative bacterium, Cupriavidus sp. strain CNP-8, AMB Express, 8 (2018) 43.

[109] C. Liedtke, T. Luedde, T. Sauerbruch, D. Scholten, K. Streetz, F. Tacke, R. Tolba, C. Trautwein, J. Trebicka, R. Weiskirchen, Experimental liver fibrosis research: update on animal models, legal issues and translational aspects, Fibrogenesis & Tissue Repair, 6 (2013) 1–25.

[110] M.H. Keshavarz, Combustible Organic Materials: Determination and Prediction of Combustion Properties, Walter de Gruyter GmbH & Co KG, (2018).

[111] M.H. Keshavarz, T.M. Klapötke, Energetic Compounds: Methods for Prediction of their Performance, Walter de Gruyter GmbH & Co KG, (2017).

[112] M.H. Keshavarz, T.M. Klapötke, T.M. Klapotke, The Properties of Energetic Materials: Sensitivity, Physical and Thermodynamic Properties, Walter de Gruyter GmbH & Co KG, (2017).

[113] S. Zeman, M. Jungová, Sensitivity and performance of energetic materials, Propellants, Explosives, Pyrotechnics, 41 (2016) 426–451.

[114] D.E. Rickert, Toxicity of nitroaromatic compounds, CRC Press, (1985).

[115] Č. Narimantas, A. Nemeikait, H. Nivinskas, Ž. Anusevičius, J. Šarlauskas, Quantitative structure–activity relationships in enzymatic single-electron reduction of nitroaromatic explosives: implications for their cytotoxicity, Biochimica et Biophysica Acta (BBA)-General Subjects, 1528 (2001) 31–38.

[116] H. Schmitt, R. Altenburger, B. Jastorff, G. Schüürmann, Quantitative structure– activity analysis of the algae toxicity of nitroaromatic compounds, Chemical Research in Toxicology, 13 (2000) 441–450.

[117] S.S. Talmage, D.M. Opresko, C.J. Maxwell, C.J. Welsh, F.M. Cretella, P.H. Reno, F.B. Daniel, Nitroaromatic munition compounds: Environmental effects and screening values, in: Reviews of environmental contamination and toxicology, Springer, (1999), pp. 1–156.

[118] X. Wang, Z. Lin, D. Yin, S. Liu, L. Wang, 2D/3D-QSAR comparative study on mutagenicity of nitroaromatics, Science in China Series B: Chemistry, 48 (2005) 246–252.

[119] O. Mekenyan, D.W. Roberts, W. Karcher, Molecular orbital parameters as predictors of skin sensitization potential of halo-and pseudohalobenzenes acting as SNAr electrophiles, Chemical Research in Toxicology, 10 (1997) 994–1000.

[120] B. Bukowska, S. Kowalska, The presence and toxicity of phenol derivatives – their effect on human erythrocytes, Current Topics in Biophysics, 27 (2003) 43–51.

[121] W.P. Cunningham, B.W. Saigo, M.A. Cunningham, Environmental science: a global concern, McGraw-Hill Boston, MA, (2001).

[122] Y. Hao, T. Fan, G. Sun, F. Li, N. Zhang, L. Zhao, R. Zhong, Environmental toxicity risk evaluation of nitroaromatic compounds: Machine learning driven binary/multiple classification and design of safe alternatives, Food and Chemical Toxicology, 170 (2022) 113461.

[123] V. Kuz'Min, E. Muratov, A. Artemenko, L. Gorb, M. Qasim, J. Leszczynski, The effect of nitroaromatics' composition on their toxicity in vivo: novel, efficient non-additive 1D QSAR analysis, Chemosphere, 72 (2008) 1373–1380.

[124] V.E. Kuz'min, E.N. Muratov, A.G. Artemenko, L. Gorb, M. Qasim, J. Leszczynski, The effects of characteristics of substituents on toxicity of the nitroaromatics: HiT QSAR study, Journal of Computer-Aided Molecular Design, 22 (2008) 747–759.

[125] V. Agrawal, P. Khadikar, QSAR prediction of toxicity of nitrobenzenes, Bioorganic & Medicinal Chemistry, 9 (2001) 3035–3040.

[126] A. Niazi, S. Jameh-Bozorghi, D. Nori-Shargh, Prediction of toxicity of nitrobenzenes using ab initio and least squares support vector machines, Journal of Hazardous Materials, 151 (2008) 603–609.

[127] U.S.N.L.o.M. TOXNET-ChemIDplus, https://chem.nlm.nih.gov/chemidplus/chemidlite.jsp, in, (2018).

[128] A. Gooch, N. Sizochenko, B. Rasulev, L. Gorb, J. Leszczynski, In vivo toxicity of nitroaromatics: a comprehensive quantitative structure–activity relationship study, Environmental Toxicology and Chemistry, 36 (2017) 2227–2233.

[129] Y. Hao, G. Sun, T. Fan, X. Sun, Y. Liu, N. Zhang, L. Zhao, R. Zhong, Y. Peng, Prediction on the mutagenicity of nitroaromatic compounds using quantum chemistry descriptors based QSAR and machine learning derived classification methods, Ecotoxicology and Environmental Safety, 186 (2019) 109822.

[130] OECD (Organization for Economic Co-Operation and Development), The Report from the Expert Group on (Quantitative) Structure-Activity Relationships [(Q)SARs] on the Principles for the Validation of (Q)SARs. 2nd Meeting of the Ad Hoc Expert Group on QSARs. OECD Headquarters, 20–21 September (2004).

[131] R. Todeschini, V. Consonni, Handbook of Molecular Descriptors, John Wiley & Sons, (2008).

[132] G.K. Jillella, K. Khan, K. Roy, Application of QSARs in identification of mutagenicity mechanisms of nitro and amino aromatic compounds against Salmonella typhimurium species, Toxicology in Vitro, 65 (2020) 104768.

[133] C.W. Yap, PaDEL-descriptor: An open source software to calculate molecular descriptors and fingerprints, Journal of Computational Chemistry, 32 (2011) 1466–1474.

[134] http://www.talete.mi.it/products/dragondescription.htm,in.

[135] http://www.yapcwsoft.com/dd/padeldescripto,in.

[136] T. Fan, G. Sun, L. Zhao, X. Cui, R. Zhong, QSAR and classification study on prediction of acute oral toxicity of N-nitroso compounds, International Journal of Molecular Sciences, 19 (2018) 3015.

[137] Y. Hao, G. Sun, T. Fan, X. Tang, J. Zhang, Y. Liu, N. Zhang, L. Zhao, R. Zhong, Y. Peng, In vivo toxicity of nitroaromatic compounds to rats: QSTR modelling and interspecies toxicity relationship with mouse, Journal of Hazardous Materials, 399 (2020) 122981.

[138] L.-L. Wang, J.-J. Ding, L. Pan, L. Fu, J.-H. Tian, D.-S. Cao, H. Jiang, X.-Q. Ding, Quantitative structure-toxicity relationship model for acute toxicity of organophosphates via multiple administration routes in rats and mice, Journal of Hazardous Materials, 401 (2021) 123724.

[139] Y.-z. Sun, Z.-j. Li, X.-l. Yan, L. Wang, F.-h. Meng, Study on the quantitative structure–toxicity relationships of benzoic acid derivatives in rats via oral LD50, Medicinal Chemistry Research, 18 (2009) 712–724.

[140] D. Mondal, K. Ghosh, A.T. Baidya, A.M. Gantait, S. Gayen, Identification of structural fingerprints for in vivo toxicity by using Monte Carlo based QSTR modeling of nitroaromatics, Toxicology Mechanisms and Methods, 30 (2020) 257–265.

[141] M. Keshavarz, A. Akbarzadeh, A simple approach for assessment of toxicity of nitroaromatic compounds without using complex descriptors and computer codes, SAR and QSAR in Environmental Research, 30 (2019) 347–361.

[142] H. Pouretedal, M. Keshavarz, Prediction of toxicity of nitroaromatic compounds through their molecular structures, Journal of the Iranian Chemical Society, 8 (2011) 78–89.

[143] A. Daghighi, G.M. Casanola-Martin, T. Timmerman, D. Milenković, B. Lučić, B. Rasulev, In silico prediction of the toxicity of nitroaromatic compounds: application of ensemble learning QSAR approach, Toxics, 10 (2022) 746.

[144] P.J. O'Brien, A.G. Siraki, N. Shangari, Aldehyde sources, metabolism, molecular toxicity mechanisms, and possible effects on human health, Critical Reviews in Toxicology, 35 (2005) 609–662.

[145] C. Gampe, V.A. Verma, Curse or cure? A perspective on the developability of aldehydes as active pharmaceutical ingredients, Journal of Medicinal Chemistry, 63 (2020) 14357–14381.

[146] T.I. Netzeva, T.W. Schultz, QSARs for the aquatic toxicity of aromatic aldehydes from Tetrahymena data, Chemosphere, 61 (2005) 1632–1643.

[147] S. Kar, A. Harding, K. Roy, P. Popelier, QSAR with quantum topological molecular similarity indices: toxicity of aromatic aldehydes to Tetrahymena pyriformis, SAR and QSAR in Environmental Research, 21 (2010) 149–168.

[148] K. Roy, R.N. Das, QSTR with extended topochemical atom (ETA) indices. 14. QSAR modeling of toxicity of aromatic aldehydes to Tetrahymena pyriformis, Journal of Hazardous Materials, 183 (2010) 913–922.

[149] M. Asadollahi-Baboli, Straightforward MIA-QSTR evaluation of environmental toxicities of aromatic aldehydes to Tetrahymena pyriformis, SAR and QSAR in Environmental Research, 24 (2013) 1041–1050.

[150] A. Ousaa, B. Elidrissi, M. Ghamali, S. Chtita, A. Aouidate, M. Bouachrine, T. Lakhlifi, Quantitative structure-toxicity relationship studies of aromatic aldehydes to Tetrahymena pyriformis based on electronic and topological descriptors, Journal of Materials and Environmental Science, 9 (2018) 256–266.

[151] B. Louis, V.K. Agrawal, QSAR modeling of aquatic toxicity of aromatic aldehydes using artificial neural network (ANN) and multiple linear regression (MLR), Journal of the Indian Chemical Society, 88 (2011) 99.

[152] W.D. Ihlenfeldt, E.E. Bolton, S.H. Bryant, The PubChem chemical structure sketcher, Journal of Cheminformatics, 1 (2009) 1–9.

[153] K.L. Linge, I. Kristiana, D. Liew, A. Holman, C.A. Joll, Halogenated semivolatile acetonitriles as chloramination disinfection by-products in water treatment: A new formation pathway from activated aromatic compounds, Environmental Science: Processes & Impacts, 22 (2020) 653–662.

[154] L.W. Tang, Y. Alias, R. Zakaria, P.M. Woi, Progress in electrochemical sensing of heavy metals based on amino acids and its composites, Critical Reviews in Analytical Chemistry, (2021) 1–18.

[155] H. Jäckel, W. Klein, Prediction of mammalian toxicity by quantitative structure activity relationships: aliphatic amines and anilines, Quantitative Structure-Activity Relationships, 10 (1991) 198–204.

[156] L. Xu, Y. Wu, C. Hu, H. Li, A QSAR of the toxicity of amino-benzenes and their structures, Science in China Series B: Chemistry, 43 (2000) 129–136.

[157] M.K. Mahani, M. Chaloosi, M.G. Maragheh, A.R. Khanchi, D. Afzali, Prediction of acute in vivo toxicity of some amine and amide drugs to rats by multiple linear regression, partial least squares and an artificial neural network, Analytical Sciences, 23 (2007) 1091–1095.

[158] L. Xu, J. Yang, Three-dimensional structural features and the toxicity of aminobenzenes and phenols, Science in China Series B: Chemistry, 46 (2003) 431–438.

[159] H.R. Pouretedal, M.H. Keshavarz, A. Abbasi, A new approach for accurate prediction of toxicity of amino compounds, Journal of the Iranian Chemical Society, 12 (2015) 487–502.

[160] G. He, L. Feng, H. Chen, A QSAR study of the acute toxicity of halogenated phenols, Procedia Engineering, 43 (2012) 204–209.

[161] T. Langer, S.D. Bryant, 3D quantitative structure–property relationships, in: W.C. Georges (Ed.) The Practice of Medicinal Chemistry, Elsevier, (2008), pp. 587–604.

[162] X.H. Chen, Z.J. Shan, H.L. Zhai, QSAR models for predicting the toxicity of halogenated phenols to Tetrahymena, Toxicological & Environmental Chemistry, 99 (2017) 273–284.

[163] G. Klebe, U. Abraham, T. Mietzner, Molecular similarity indices in a comparative analysis (CoMSIA) of drug molecules to correlate and predict their biological activity, Journal of Medicinal Chemistry, 37 (1994) 4130–4146.

[164] Y. Wang, H. Liu, X. Yang, L. Wang, Aquatic toxicity and aquatic ecological risk assessment of wastewater-derived halogenated phenolic disinfection byproducts, Science of the Total Environment, 809 (2022) 151089.

[165] J.M. Bermúdez-Saldaña, M.T. Cronin, Quantitative structure–activity relationships for the toxicity of organophosphorus and carbamate pesticides to the Rainbow trout Onchorhyncus mykiss, Pest Management Science: formerly Pesticide Science, 62 (2006) 819–831.

[166] S.A. Senior, M.D. Madbouly, QSTR of the toxicity of some organophosphorus compounds by using the quantum chemical and topological descriptors, Chemosphere, 85 (2011) 7–12.

[167] J. Zhao, S. Yu, Quantitative structure–activity relationship of organophosphate compounds based on molecular interaction fields descriptors, Environmental Toxicology and Pharmacology, 35 (2013) 228–234.

[168] A. Can, Quantitative structure–toxicity relationship (QSTR) studies on the organophosphate insecticides, Toxicology Letters, 230 (2014) 434–443.

[169] X. Ding, J. Ding, D. Li, L. Pan, C. Pei, Toxicity prediction of organophosphorus chemical reactivity compounds based on conceptual DFT, Acta Physico-Chimica Sinica, 34 (2018) 314–322.

[170] R.L. Camacho-Mendoza, E. Aquino-Torres, V. Cordero-Pensado, J. Cruz-Borbolla, J.G. Alvarado-Rodríguez, P. Thangarasu, C.Z. Gómez-Castro, A new computational model for the prediction of toxicity of phosphonate derivatives using QSPR, Molecular Diversity, 22 (2018) 269–280.

[171] M. Kianpour, E. Mohammadinasab, T.M. Isfahani, Prediction of oral acute toxicity of organophosphates using QSAR methods, Current Computer-Aided Drug Design, 17 (2021) 38–56.

[172] N. Basant, S. Gupta, K.P. Singh, Modeling the toxicity of chemical pesticides in multiple test species using local and global QSTR approaches, Toxicology Research, 5 (2016) 340–353.

[173] R.N. Das, K. Roy, P.L. Popelier, Interspecies quantitative structure–toxicity–toxicity (QSTTR) relationship modeling of ionic liquids. Toxicity of ionic liquids to V. fischeri, D. magna and S. vacuolatus, Ecotoxicology and Environmental Safety, 122 (2015) 497–520.

[174] S. Cassani, S. Kovarich, E. Papa, P.P. Roy, L. van der Wal, P. Gramatica, Daphnia and fish toxicity of (benzo) triazoles: Validated QSAR models, and interspecies quantitative activity–activity modelling, Journal of Hazardous Materials, 258 (2013) 50–60.

[175] G. Ilia, A. Borota, S. Funar-Timofei, Interspecies quantitative structure-toxicity-toxicity relationships for predicting the acute toxicity of organophosphorous compounds, Chemistry Proceedings, 8 (2021) 32.

[176] D.D. Nguyen, Review on polychlorinated naphthalenes (PCNs): properties, sources, characteristics of emission and atmospheric level, Journal of Technical Education Science, (2020) 1–13.

[177] C.I.C.A.D.C.N.W. World Health Organization, Geneva. http://apps.who.int/iris/bitstream/handle/10665/42403/9241530340.pdf?sequence=1,in.

[178] J. Falandysz, Polychlorinated naphthalenes: an environmental update, Environmental Pollution, 101 (1998) 77–90.

[179] A. Nath, P.K. Ojha, K. Roy, Computational modeling of aquatic toxicity of polychlorinated naphthalenes (PCNs) employing 2D-QSAR and chemical read-across, Aquatic Toxicology, 257 (2023) 106429.

[180] L.B. Kier, L.H. Hall, Molecular connectivity in structure-activity analysis, Research Studies, (1986).

[181] K. Naumann, Influence of chlorine substituents on biological activity of chemicals: a review, in, Wiley Online Library, (2000).

[182] V. Rastija, M. Medić-Šarić, QSAR study of antioxidant activity of wine polyphenols, European Journal of Medicinal Chemistry, 44 (2009) 400–408.

[183] F. De Logu, P. Geppetti, Ion Channel Pharmacology for Pain Modulation, in: J. Barrett, C. Page, M. Michel (Eds.) Concepts and Principles of Pharmacology. Handbook of Experimental Pharmacology, Springer, Cham, (2019).

[184] M.S. Alavi, A. Shamsizadeh, G. Karimi, A. Roohbakhsh, Transient receptor potential ankyrin 1 (TRPA1)-mediated toxicity: friend or foe? Toxicology Mechanisms and Methods, 30 (2019) 1–18.

[185] C.T. de David Antoniazzi, S.D.-T. De Prá, P.R. Ferro, M.A. Silva, G. Adamante, A.S. de Almeida, C. Camponogara, C.R. da Silva, G. de Bem Silveira, P.C.L. Silveira, Topical treatment with a transient receptor potential ankyrin 1 (TRPA1) antagonist reduced nociception and inflammation in a thermal lesion model in rats, European Journal of Pharmaceutical Sciences, 125 (2018) 28–38.

[186] X. Zheng, Y. Tai, D. He, B. Liu, C. Wang, X. Shao, S.-E. Jordt, B. Liu, ETAR and protein kinase A pathway mediate ET-1 sensitization of TRPA1 channel: A molecular mechanism of ET-1-induced mechanical hyperalgesia, Molecular Pain, 15 (2019) 1–11.

[187] M.S. Alavi, A. Shamsizadeh, G. Karimi, A. Roohbakhsh, Transient receptor potential ankyrin 1 (TRPA1)-mediated toxicity: friend or foe? Toxicology Mechanisms and Methods, (2019) 1–18.

[188] A. Jha, P. Sharma, V. Anaparti, M.H. Ryu, A.J. Halayko, A role for transient receptor potential ankyrin 1 cation channel (TRPA1) in airway hyper-responsiveness? Canadian Journal of Physiology and Pharmacology, 93 (2015) 171–176.

[189] S. Achanta, S.-E. Jordt, TRPA1: Acrolein meets its target, Toxicology and Applied Pharmacology, 324 (2017) 45.

[190] T. Voets, J. Vriens, R. Vennekens, Targeting TRP channels–valuable alternatives to combat pain, lower urinary tract disorders, and type 2 diabetes? Trends in Pharmacological Sciences, (2019).

[191] T.I. Kichko, W. Neuhuber, G. Kobal, P.W. Reeh, The roles of TRPV1, TRPA1 and TRPM8 channels in chemical and thermal sensitivity of the mouse oral mucosa, European Journal of Neuroscience, 47 (2018) 201–210.

[192] E. Correa, W. Quiñones, F. Echeverri, Methyl-N-methylanthranilate, a pungent compound from Citrus reticulata Blanco leaves, Pharmaceutical Biology, 54 (2016) 569–571.

[193] J.M. McPartland, M. Duncan, V. Di Marzo, R.G. Pertwee, Are cannabidiol and Δ9-tetrahydrocannabivarin negative modulators of the endocannabinoid system? A systematic review, British Journal of Pharmacology, 172 (2015) 737–753.

[194] D. Kim, M.-H. Lee, S.K. Kim, Involvement of TRPA1 in the cinnamaldehyde-induced pulpal blood flow change in the feline dental pulp, Restorative Dentistry & Endodontics, 41 (2016) 202–209.

[195] Y.A. Alpizar, B. Boonen, M. Gees, P. Uvin, T. Voets, D. De Ridder, W. Everaerts, K. Talavera, TRPV1 Contributes to Acrolein-Induced Toxicity, Biophysical Journal, 112 (2017) 410a.

[196] G. Pozsgai, I.Z. Bátai, E. Pintér, Effects of sulfide and polysulfides transmitted by direct or signal transduction-mediated activation of TRPA1 channels, British Journal of Pharmacology, 176 (2019) 628–645.

[197] M.V.S. Suryanarayana, A.K. Nigam, A. Mazumder, P.K. Gutch, Studies on thermal degradation of riot control agent dibenz [b, f]-1, 4-oxazepine (CR), Indian Journal of Chemistry – Section B (IJC-B) 56B (2017) 862–871.

[198] P.K. Bahia, T.A. Parks, K.R. Stanford, D.A. Mitchell, S. Varma, S.M. Stevens, T.E. Taylor-Clark, The exceptionally high reactivity of Cys 621 is critical for electrophilic activation of the sensory nerve ion channel TRPA1, The Journal of General Physiology, 147 (2016) 451–465.

[199] M.J. Ko, L.C. Ganzen, E. Coskun, A.A. Mukadam, Y.F. Leung, R.M. van Rijn, A critical evaluation of TRPA1-mediated locomotor behavior in zebrafish as a screening tool for novel anti-nociceptive drug discovery, Scientific Reports, 9 (2019) 2430.

[200] Y. Ai, F.-J. Song, S.-T. Wang, Q. Sun, P.-H. Sun, Molecular modeling studies on 11H-dibenz[b, e] azepine and dibenz[b, f][1, 4]oxazepine derivatives as potent agonists of the human TRPA1 receptor, Molecules, 15 (2010) 9364–9379.

[201] I. Novak, L. Klasinc, S.P. McGlynn, Electronic structure of 11H-dibenz(b, f)azepines, Journal of Electron Spectroscopy and Related Phenomena, 212 (2016) 56–61.

[202] M.H. Keshavarz, H. Fakhraian, N. Saedi, A simple model for the assessment of the agonistic activity of dibenzazepine derivatives by molecular moieties, Medicinal Chemistry Research, 30 (2021) 215–225.

[203] H.J.M. Gijsen, D. Berthelot, M. Zaja, B. Brone, I. Geuens, M. Mercken, Analogues of morphanthridine and the tear gas dibenz[b,f][1,4]oxazepine (CR) as extremely potent activators of the human transient receptor potential ankyrin 1 (TRPA1) channel, Journal of Medicinal Chemistry, 53 (2010) 7011–7020.

[204] O. Idowu, K.T. Semple, K. Ramadass, W. O'Connor, P. Hansbro, P. Thavamani, Beyond the obvious: Environmental health implications of polar polycyclic aromatic hydrocarbons, Environment International, 123 (2019) 543–557.

[205] C. Achten, J.T. Andersson, Overview of polycyclic aromatic compounds (PAC), Polycyclic Aromatic Compounds, 35 (2015) 177–186.

[206] H. Ha, K. Park, G. Kang, S. Lee, QSAR study using acute toxicity of Daphnia magna and Hyalella azteca through exposure to polycyclic aromatic hydrocarbons (PAHs), Ecotoxicology, 28 (2019) 333–342.

[207] F. Li, G. Sun, T. Fan, N. Zhang, L. Zhao, R. Zhong, Y. Peng, Ecotoxicological QSAR modelling of the acute toxicity of fused and non-fused polycyclic aromatic hydrocarbons (FNFPAHs) against two aquatic organisms: Consensus modelling and comparison with ECOSAR, Aquatic Toxicology, (2023) 106393.

[208] G. Sun, Y. Zhang, L. Pei, Y. Lou, Y. Mu, J. Yun, F. Li, Y. Wang, Z. Hao, S. Xi, Chemometric QSAR modeling of acute oral toxicity of Polycyclic Aromatic Hydrocarbons (PAHs) to rat using simple 2D descriptors and interspecies toxicity modeling with mouse, Ecotoxicology and Environmental Safety, 222 (2021) 112525.

[209] H. Huang, X. Wang, W. Ou, J. Zhao, Y. Shao, L. Wang, Acute toxicity of benzene derivatives to the tadpoles (Rana japonica) and QSAR analyses, Chemosphere, 53 (2003) 963–970.

[210] Y. Martínez-López, S.J. Barigye, O. Martínez-Santiago, Y. Marrero-Ponce, J. Green, J.A. Castillo-Garit, Prediction of aquatic toxicity of benzene derivatives using molecular descriptor from atomic weighted vectors, Environmental Toxicology and Pharmacology, 56 (2017) 314–321.

[211] B. Zoeteman, K. Harmsen, J. Linders, C. Morra, W. Slooff, Persistent organic pollutants in river water and ground water of the Netherlands, Chemosphere, 9 (1980) 231–249.

[212] D.-D. Wang, L.-L. Feng, G.-Y. He, H.-Q. Chen, QSAR studies for the acute toxicity of nitrobenzenes to the Tetrahymena pyriformis, Journal of the Serbian Chemical Society, 79 (2014) 1111–1125.

[213] J. Dan, Z. Jianguo, L. Na, R. Kaifeng, L. Xiao, H. Yi, M. Mei, Quantitative structure-activity relationships between acute toxicity of organophosphates and Vibrio qinghaiensis sp.-Q67, Asian Journal of Ecotoxicology, (2014) 71–80.

[214] M. Salahinejad, J. Ghasemi, 3D-QSAR studies on the toxicity of substituted benzenes to Tetrahymena pyriformis: CoMFA, CoMSIA and VolSurf approaches, Ecotoxicology and Environmental Safety, 105 (2014) 128–134.

[215] A. Singh, S. Kumar, A. Kapoor, P. Kumar, A. Kumar, Development of reliable quantitative structure–toxicity relationship models for toxicity prediction of benzene derivatives using semiempirical descriptors, Toxicology Mechanisms and Methods, (2022) 1–11.

[216] T. Janssens, D. Giesen, J. Mariën, N. van Straalen, C. van Gestel, D. Roelofs, Narcotic mechanisms of acute toxicity of chlorinated anilines in Folsomia candida (Collembola) revealed by gene expression analysis, Environment International, 37 (2011) 929–939.

[217] R. Todeschini, V. Consonni, Molecular descriptors for chemoinformatics. 1. Alphabetical listing, Wiley-VCH, (2009).

[218] Y. Marrero-Ponce, R. Medina-Marrero, F. Torrens, Y. Martinez, V. Romero-Zaldivar, E.A. Castro, Atom, atom-type, and total nonstochastic and stochastic quadratic fingerprints: a promising approach for modeling of antibacterial activity, Bioorganic & Medicinal Chemistry, 13 (2005) 2881–2899.

[219] J.A. Castillo-Garit, Y. Marrero-Ponce, J. Escobar, F. Torrens, R. Rotondo, A novel approach to predict aquatic toxicity from molecular structure, Chemosphere, 73 (2008) 415–427.

[220] L. Pauling, The Nature of the Chemical Bond, Cornell University Press, Ithaca, NY, (1941).

[221] R. Todeschini, P. Gramatica, New 3D molecular descriptors: the WHIM theory and QSAR applications, in: 3D QSAR in Drug Design, Springer, (2002), pp. 355–380.

[222] V. Consonni, R. Todeschini, M. Pavan, Structure/response correlations and similarity/diversity analysis by GETAWAY descriptors. 1. Theory of the novel 3D molecular descriptors, Journal of Chemical Information and Computer Sciences, 42 (2002) 682–692.

[223] Y.H. Zhao, X. Yuan, L.M. Su, W.C. Qin, M.H. Abraham, Classification of toxicity of phenols to Tetrahymena pyriformis and subsequent derivation of QSARs from hydrophobic, ionization and electronic parameters, Chemosphere, 75 (2009) 866–871.

[224] B. Hemmateenejad, A.R. Mehdipour, R. Miri, M. Shamsipur, Comparative qsar studies on toxicity of phenol derivatives using quantum topological molecular similarity indices, Chemical Biology & Drug Design, 75 (2010) 521–531.

[225] F. Abbasitabar, V. Zare-Shahabadi, In silico prediction of toxicity of phenols to Tetrahymena pyriformis by using genetic algorithm and decision tree-based modeling approach, Chemosphere, 172 (2017) 249–259.

[226] A. Del Olmo, J. Calzada, M. Nuñez, Benzoic acid and its derivatives as naturally occurring compounds in foods and as additives: Uses, exposure, and controversy, Critical Reviews in Food Science and Nutrition, 57 (2017) 3084–3103.

[227] X. Huang, J. He, X. Yan, Q. Hong, K. Chen, Q. He, L. Zhang, X. Liu, S. Chuang, S. Li, Microbial catabolism of chemical herbicides: microbial resources, metabolic pathways and catabolic genes, Pesticide Biochemistry and Physiology, 143 (2017) 272–297.

[228] Y. Kamaya, Y. Fukaya, K. Suzuki, Acute toxicity of benzoic acids to the crustacean Daphnia magna, Chemosphere, 59 (2005) 255–261.

[229] P.Y. Lee, C.Y. Chen, Toxicity and quantitative structure–activity relationships of benzoic acids to Pseudokirchneriella subcapitata, Journal of Hazardous Materials, 165 (2009) 156–161.

[230] Z. Li, Y. Sun, X. Yan, F. Meng, Study on QSTR of benzoic acid compounds with MCI, International Journal of Molecular Sciences, 11 (2010) 1228–1235.

[231] M.H. Keshavarz, F. Gharagheizi, A. Shokrolahi, S. Zakinejad, Accurate prediction of the toxicity of benzoic acid compounds in mice via oral without using any computer codes, Journal of Hazardous Materials, 237 (2012) 79–101.

[232] M.H. Keshavarz, H.R. Pouretedal, Simple and reliable prediction of toxicological activities of benzoic acid derivatives without using any experimental data or computer codes, Medicinal Chemistry Research, 22 (2013) 1238–1257.

[233] https://pubchem.ncbi.nlm.nih.gov/.

[234] P. Büscher, G. Cecchi, V. Jamonneau, G. Priotto, Human African trypanosomiasis, The Lancet, 390 (2017) 2397–2409.

[235] V. Jamonneau, P. Truc, P. Grébaut, S. Herder, S. Ravel, P. Solano, T. De Meeus, Trypanosoma brucei gambiense Group 2: the unusual suspect, Trends in Parasitology, 35 (2019) 983–995.

[236] A. Aguilar, T. Twardowski, R. Wohlgemuth, Bioeconomy for sustainable development, Biotechnology Journal, 14 (2019) 1800638.

[237] K. Possart, F.C. Herrmann, J. Jose, M.P. Costi, T.J. Schmidt, Sesquiterpene lactones with dual inhibitory activity against the Trypanosoma brucei pteridine reductase 1 and dihydrofolate reductase, Molecules, 27 (2021) 149.

[238] T.J. Schmidt, A.M. Nour, S.A. Khalid, M. Kaiser, R. Brun, Quantitative structure–antiprotozoal activity relationships of sesquiterpene lactones, Molecules, 14 (2009) 2062–2076.

[239] T.J. Schmidt, F.B. Da Costa, N.P. Lopes, M. Kaiser, R. Brun, In silico prediction and experimental evaluation of furanoheliangolide sesquiterpene lactones as potent agents against Trypanosoma brucei rhodesiense, Antimicrobial agents and chemotherapy, 58 (2014) 325–332.

[240] G.H. Trossini, V.G. Maltarollo, T.J. Schmidt, Hologram QSAR studies of antiprotozoal activities of sesquiterpene lactones, Molecules, 19 (2014) 10546–10562.

[241] N.M. Kimani, J.C. Matasyoh, M. Kaiser, M.S. Nogueira, G.H. Trossini, T.J. Schmidt, Complementary quantitative structure–activity relationship models for the antitrypanosomal activity of sesquiterpene lactones, International Journal of Molecular Sciences, 19 (2018) 3721.

[242] M.H. Keshavarz, Z. Shirazi, F. Sayehvand, A novel approach for assessment of antitrypanosomal activity of sesquiterpene lactones through additive and non-additive molecular structure parameters, Molecular Diversity, (2022) 1–10.

[243] S. Biswas, V. Thakur, P. Kaur, A. Khan, S. Kulshrestha, P. Kumar, Blood clots in COVID-19 patients: Simplifying the curious mystery, Medical Hypotheses, 146 (2021) 110371.

[244] C. Agrati, V. Mazzotta, C. Pinnetti, G. Biava, M. Bibas, Venous thromboembolism in people living with HIV infection (PWH), Translational Research, (2020).

[245] J. Huang, W. Song, H. Hua, X. Yin, F. Huang, R.N. Alolga, Antithrombotic and anticoagulant effects of a novel protein isolated from the venom of the Deinagkistrodon acutus snake, Biomedicine & Pharmacotherapy, 138 (2021) 111527.

[246] N.M. Lancy Morries, J.J. Joseph, S. Abraham, Newer oral anticoagulant therapy–prospects and practices: a review, Indian Journal of Pharmacy Practice, 13 (2020) 215.

[247] Z.-G. Sun, J.-M. Zhang, S.-C. Cui, Z.-G. Zhang, H.-L. Zhu, The Research Progress of Direct Thrombin Inhibitors, Mini Reviews in Medicinal Chemistry, 20 (2020) 1574–1585.

[248] G. Lippi, R. Gosselin, E.J. Favaloro, Current and emerging direct oral anticoagulants: state-of-the-art, in: Seminars in Thrombosis and Hemostasis, Thieme Medical Publishers, (2019), pp. 490–501.

[249] A. Dixit, S.K. Kashaw, S. Gaur, A.K. Saxena, Development of CoMFA, advance CoMFA and CoMSIA models in pyrroloquinazolines as thrombin receptor antagonist, Bioorganic & Medicinal Chemistry, 12 (2004) 3591–3598.

[250] K. Mena-Ulecia, W. Tiznado, J. Caballero, Study of the differential activity of thrombin inhibitors using docking, QSAR, molecular dynamics, and MM-GBSA, PLoS One, 10 (2015) e0142774.

[251] M. Nilsson, M. Hämäläinen, M. Ivarsson, J. Gottfries, Y. Xue, S. Hansson, R. Isaksson, T. Fex, Compounds binding to the S2– S3 pockets of thrombin, Journal of Medicinal Chemistry, 52 (2009) 2708–2715.

[252] S.S. Bhunia, K.K. Roy, A.K. Saxena, Profiling the structural determinants for the selectivity of representative factor-Xa and thrombin inhibitors using combined ligand-based and structure-based approaches, Journal of Chemical Information and Modeling, 51 (2011) 1966–1985.

[253] M.H. Keshavarz, Z. Shirazi, M. Mohajeri, Simple method for assessment of activities of thrombin inhibitors through their molecular structure parameters, Computers in Biology and Medicine, 146 (2022) 105640.

[254] D.E. Nichols, Chemistry and structure-activity relationships of psychedelics, in: A.L. Halberstadt, F.X. Vollenweider, D.E. Nichols (Eds.) Behavioral Neurobiology of Psychedelic Drugs, Springer, Springer-Verlag GmbH Germany, (2017), pp. 1–43.

[255] M. Thakur, A. Thakur, P.V. Khadikar, QSAR studies on psychotomimetic phenylalkylamines, Bioorganic & Medicinal Chemistry, 12 (2004) 825–831.

[256] A. Aouidate, A. Ghaleb, M. Ghamali, S. Chtita, M. Choukrad, A. Sbai, M. Bouachrine, T. Lakhlifi, Combining DFT and QSAR studies for predicting psychotomimetic activity of substituted phenethylamines using statistical methods, Journal of Taibah University for Science, 10 (2016) 787–796.

[257] B.W. Clare, The frontier orbital phase angles: novel QSAR descriptors for benzene derivatives, applied to phenylalkylamine hallucinogens, Journal of Medicinal Chemistry, 41 (1998) 3845–3856.

[258] M.J.-M. Takač, J.D.C. Magina, T. Takač, Evaluation of phenylethylamine type entactogens and their metabolites relevant to ecotoxicology–a QSAR study, Acta Pharmaceutica, 69 (2019) 563–584.

[259] M.H. Keshavarz, Z. Shirazi, M.A. Rezayat, A simple method for assessing the psychotomimetic activity of the substituted phenethylamines, Zeitschrift für anorganische und allgemeine Chemie, 647 (2021) 651–662.

[260] A. Shulgin, A. Shulgin, PIHKAL: A Chemical Love Story, Transform Press, Berkeley, CA 94701, (1991).

[261] T. Rengarajan, P. Rajendran, N. Nandakumar, B. Lokeshkumar, P. Rajendran, I. Nishigaki, Exposure to polycyclic aromatic hydrocarbons with special focus on cancer, Asian Pacific Journal of Tropical Biomedicine, 5 (2015) 182–189.

[262] K. Sun, Y. Song, F. He, M. Jing, J. Tang, R. Liu, A review of human and animals exposure to polycyclic aromatic hydrocarbons: Health risk and adverse effects, photo-induced toxicity and regulating effect of microplastics, Science of the Total Environment, 773 (2021) 145403.

[263] A. Kumar, T. Podder, V. Kumar, P.K. Ojha, Risk assessment of aromatic organic chemicals to T. pyriformis in environmental protection using regression-based QSTR and Read-Across algorithm, Process Safety and Environmental Protection, (2022).

[264] D.V. Parke, The Biochemistry of Foreign Compounds: International Series of Monographs in Pure and Applied Biology: Biochemistry, Elsevier, (2013).

[265] A. Nath, K. Roy, Chemometric modeling of acute toxicity of diverse aromatic compounds against Rana japonica, Toxicology in Vitro, 83 (2022) 105427.

[266] Y. Yun, A. Edginton, Correlation-based prediction of tissue-to-plasma partition coefficients using readily available input parameters, Xenobiotica, 43 (2013) 839–852.

[267] R.M. LoPachin, T. Gavin, Molecular mechanisms of aldehyde toxicity: a chemical perspective, Chemical research in toxicology, 27 (2014) 1081–1091.

[268] K. McDonough, Amphibian species of the world: an online reference (version 6), Reference Reviews, 28 (2014) 32–32.

[269] D.S. Bower, L.A. Brannelly, C.A. McDonald, R.J. Webb, S.E. Greenspan, M. Vickers, M.G. Gardner, M.J. Greenlees, A review of the role of parasites in the ecology of reptiles and amphibians, Austral Ecology, 44 (2019) 433–448.

[270] E.P.o.P.P. Products, t. Residues, C. Ockleford, P. Adriaanse, P. Berny, T. Brock, S. Duquesne, S. Grilli, A.F. Hernandez-Jerez, S.H. Bennekou, M. Klein, Scientific Opinion on the state of the science on pesticide risk assessment for amphibians and reptiles, EFSA Journal, 16 (2018) e05125.

[271] E. Carnesecchi, C. Toma, A. Roncaglioni, N. Kramer, E. Benfenati, J.L.C. Dorne, Integrating QSAR models predicting acute contact toxicity and mode of action profiling in honey bees (A. mellifera): Data curation using open source databases, performance testing and validation, Science of the Total Environment, 735 (2020) 139243.

[272] G.J. Lavado, D. Baderna, E. Carnesecchi, A.P. Toropova, A.A. Toropov, J.L.C. Dorne, E. Benfenati, QSAR models for soil ecotoxicity: Development and validation of models to predict reproductive toxicity of organic chemicals in the collembola Folsomia candida, Journal of Hazardous Materials, 423 (2022) 127236.

[273] T. Austin, M. Denoyelle, A. Chaudry, S. Stradling, C. Eadsforth, European Chemicals Agency dossier submissions as an experimental data source: Refinement of a fish toxicity model for predicting acute LC50 values, Environmental Toxicology and Chemistry, 34 (2015) 369–378.

[274] V.K. Agrawal, S. Chaturvedi, M.H. Abraham, P.V. Khadikar, QSAR Study on tadpole narcosis, Bioorganic & Medicinal Chemistry, 11 (2003) 4523–4533.

[275] M. Jaiswal, P. Khadikar, QSAR study on tadpole narcosis using PI index: a case of heterogenous set of compounds, Bioorganic & Medicinal Chemistry, 12 (2004) 1731–1736.

[276] S. Sahoo, C. Adhikari, M. Kuanar, B. K Mishra, A short review of the generation of molecular descriptors and their applications in quantitative structure property/activity relationships, Current Computer-Aided Drug Design, 12 (2016) 181–205.

[277] C. Adhikari, Quantitative structure-activity relationships of aquatic narcosis: a review, Current Computer-Aided Drug Design, 14 (2018) 7–28.

[278] S. Wang, L.C. Yan, S.S. Zheng, T.T. Li, L.Y. Fan, T. Huang, C. Li, Y.H. Zhao, Toxicity of some prevalent organic chemicals to tadpoles and comparison with toxicity to fish based on mode of toxic action, Ecotoxicology and Environmental Safety, 167 (2019) 138–145.

[279] L. Wang, P. Xing, C. Wang, X. Zhou, Z. Dai, L. Bai, Maximal information coefficient and support vector regression based nonlinear feature selection and QSAR modeling on toxicity of alcohol compounds to tadpoles of rana temporaria, Journal of the Brazilian Chemical Society, 30 (2019) 279–285.

[280] A.A. Toropov, M.R. Di Nicola, A.P. Toropova, A. Roncaglioni, E. Carnesecchi, N.I. Kramer, A.J. Williams, M.E. Ortiz-Santaliestra, E. Benfenati, J.-L.C. Dorne, A regression-based QSAR-model to predict acute toxicity of aromatic chemicals in tadpoles of the Japanese brown frog (Rana japonica): Calibration, validation, and future developments to support risk assessment of chemicals in amphibians, Science of the Total Environment, 830 (2022) 154795.

[281] S. Ahmadi, S. Lotfi, P. Kumar, Quantitative structure–toxicity relationship models for predication of toxicity of ionic liquids toward leukemia rat cell line IPC-81 based on index of ideality of correlation, Toxicology Mechanisms and Methods, 32 (2022) 302–312.

[282] D. Weininger, SMILES, a chemical language and information system. 1. Introduction to methodology and encoding rules, Journal of Chemical Information and Computer Sciences, 28 (1988) 31–36.

[283] A. Nath, P. De, K. Roy, QSAR modelling of inhalation toxicity of diverse volatile organic molecules using no observed adverse effect concentration (NOAEC) as the endpoint, Chemosphere, 287 (2022) 131954.

[284] R.A. Gupta, S.G. Kaskhedikar, Synthesis, antitubercular activity, and QSAR analysis of substituted nitroaryl analogs: chalcone, pyrazole, isoxazole, and pyrimidines, Medicinal Chemistry Research, 22 (2013) 3863–3880.

[285] S. Luo, B. Wu, X. Xiong, J. Wang, Short-term toxicity of ammonia, nitrite, and nitrate to early life stages of the rare minnow (Gobiocypris rarus), Environmental Toxicology and Chemistry, 35 (2016) 1422–1427.

[286] J.-H. Kim, Y.J. Kang, K.I. Kim, S.K. Kim, J.-H. Kim, Toxic effects of nitrogenous compounds (ammonia, nitrite, and nitrate) on acute toxicity and antioxidant responses of juvenile olive flounder, Paralichthys olivaceus, Environmental Toxicology and Pharmacology, 67 (2019) 73–78.

[287] V. Consonni, R. Todeschini, Molecular Descriptors for Chemoinformatics: Volume I: Alphabetical Listing/Volume II: Appendices, References, John Wiley & Sons, (2009).

[288] M. Grzonkowska, A. Sosnowska, M. Barycki, A. Rybinska, T. Puzyn, How the structure of ionic liquid affects its toxicity to Vibrio fischeri? Chemosphere, 159 (2016) 199–207.

[289] F. Luan, T. Wang, L. Tang, S. Zhang, M.N.D.S. Cordeiro, Estimation of the toxicity of different substituted aromatic compounds to the aquatic ciliate tetrahymena pyriformis by QSAR approach, Molecules, 23 (2018) 1002.

[290] A. Sannigrahi, AB initio molecular orbital calculations of bond index and valency, Advances in Quantum Chemistry, 23 (1992) 301–351.

[291] A. Katritzky, V. Lobanov, M. Karelson, Comprehensive Descriptors for Structural and Statistical Analysis, University of Florida Gainsville, FL, USA, (1994).

[292] O. Štrouf, Chemical Pattern Recognition, Research Studies Press, Baldock, UK, (1986).

[293] M. Van der Perk, Soil and Water Contamination 2nd ed., CRC Press, Taylor and Francis, London, UK, (2013).

[294] S.A. Fast, V.G. Gude, D.D. Truax, J. Martin, B.S. Magbanua, A critical evaluation of advanced oxidation processes for emerging contaminants removal, Environmental Processes, 4 (2017) 283–302.

[295] M. Cvetnic, D.J. Perisic, M. Kovacic, S. Ukic, T. Bolanca, B. Rasulev, H. Kusic, A.L. Bozic, Toxicity of aromatic pollutants and photooxidative intermediates in water: a QSAR study, Ecotoxicology and Environmental Safety, 169 (2019) 918–927.

[296] T. Garland, A.C. Barr, Toxic plants and other natural toxicants, Cabi Publishing, New York, USA, (1998).

[297] M.T. Cronin, Predicting Chemical Toxicity and Fate, CRC Press, Boca Raton, USA, (2004).

[298] R. Todeschini, V. Consonni, Descriptors from Molecular Geometry, in: J. Gasteiger (Ed.) Handbook of Chemoinformatics, Wiley-VCH Verlag GmbH, Weinheim, (2003), pp. 1004–1033.

[299] G. Melagraki, A. Afantitis, H. Sarimveis, O. Igglessi-Markopoulou, A. Alexandridis, A novel RBF neural network training methodology to predict toxicity to Vibrio fischeri, Molecular Diversity, 10 (2006) 213–221.

[300] K. Roy, I. Mitra, Electrotopological state atom (E-state) index in drug design, QSAR, property prediction and toxicity assessment, Current Computer-Aided Drug Design, 8 (2012) 135–158.

[301] J. Huuskonen, QSAR modeling with the electrotopological state indices: predicting the toxicity of organic chemicals, Chemosphere, 50 (2003) 949–953.

[302] D. Juretic, H. Kusic, D.D. Dionysiou, B. Rasulev, A.L. Bozic, Modeling of photooxidative degradation of aromatics in water matrix; combination of mechanistic and structural-relationship approach, Chemical Engineering Journal, 257 (2014) 229–241.

[303] P. Cañizares, C. Saez, J. Lobato, M.A. Rodrigo, Detoxification of synthetic industrial wastewaters using electrochemical oxidation with boron-doped diamond anodes, Journal of Chemical Technology & Biotechnology, 81 (2006) 352–358.

[304] D. Juretic, J. Puric, H. Kusic, V. Marin, A.L. Bozic, Structural influence on photooxidative degradation of halogenated phenols, Water, Air, & Soil Pollution, 225 (2014) 1–18.

[305] T. Ye, Z. Wei, R. Spinney, C.-J. Tang, S. Luo, R. Xiao, D.D. Dionysiou, Chemical structure-based predictive model for the oxidation of trace organic contaminants by sulfate radical, Water Research, 116 (2017) 106–115.

[306] L.H. Hall, B. Mohney, L.B. Kier, The electrotopological state: structure information at the atomic level for molecular graphs, Journal of Chemical Information and Computer Sciences, 31 (1991) 76–82.

[307] R.N. Das, T.E. Sintra, J.A. Coutinho, S.P. Ventura, K. Roy, P.L. Popelier, Development of predictive QSAR models for Vibrio fischeri toxicity of ionic liquids and their true external and experimental validation tests, Toxicology Research, 5 (2016) 1388–1399.

[308] Q. Su, W. Lu, D. Du, F. Chen, B. Niu, K.-C. Chou, Prediction of the aquatic toxicity of aromatic compounds to tetrahymena pyriformis through support vector regression, Oncotarget, 8 (2017) 49359.

[309] A.K. Moorthy, B.G. Rathi, S.P. Shukla, K. Kumar, V.S. Bharti, Acute toxicity of textile dye Methylene blue on growth and metabolism of selected freshwater microalgae, Environmental Toxicology Pharmacology, 82 (2021) 103552.

[310] M. Singh, R. Singh, P. Mishra, R. Sengar, U. Shahi, In-vitro compatibility of Trichoderma harzianum with systemic fungicides, International Journal of Childhood and Society, 9 (2021) 2884–2888.

[311] R. Maurya, A.K. Pandey, Importance of protozoa Tetrahymena in toxicological studies: A review, Science of The Total Environment, 741 (2020) 140058.

[312] K.V. Neethu, S.B. Nandan, N.D.D. Xavier, P.R. Jayachandran, P.R. Anu, A.M. Midhun, D. Mohan, S.R. Marigoudar, A multibiomarker approach to assess lead toxicity on the black clam, Villorita cyprinoides (Gray, 1825), from Cochin estuarine system (CES), southwest coast, India, Environmental Science and Pollution Research, 28 (2021) 1775–1788.

[313] T. Austin, C. Eadsforth, Development of a chronic fish toxicity model for predicting sub-lethal NOEC values for non-polar narcotics, SAR and QSAR in Environmental Research, 25 (2014) 147–160.

[314] G. Tugcu, M.T. Saçan, A multipronged QSAR approach to predict algal low-toxic-effect concentrations of substituted phenols and anilines, Journal of Hazardous Materials, 344 (2018) 893–901.

[315] M.T. Cronin, (Q) SARs to predict environmental toxicities: current status and future needs, Environmental Science: Processes & Impacts, 19 (2017) 213–220.

[316] A. Seth, K. Roy, QSAR modeling of algal low level toxicity values of different phenol and aniline derivatives using 2D descriptors, Aquatic Toxicology, 228 (2020) 105627.

[317] W.-E. Liu, Z. Chen, L.-P. Yang, H.Y. Au-Yeung, W. Jiang, Molecular recognition of organophosphorus compounds in water and inhibition of their toxicity to acetylcholinesterase, Chemical Communications, 55 (2019) 9797–9800.

[318] P. Kovacic, R. Somanathan, Nitroaromatic compounds: Environmental toxicity, carcinogenicity, mutagenicity, therapy and mechanism, Journal of Applied Toxicology, 34 (2014) 810–824.

[319] H. Zhu, A. Tropsha, D. Fourches, A. Varnek, E. Papa, P. Gramatica, T. Oberg, P. Dao, A. Cherkasov, I.V. Tetko, Combinatorial QSAR modeling of chemical toxicants tested against Tetrahymena pyriformis, Journal of Chemical Information and Modeling, 48 (2008) 766–784.

[320] X. Yu, Y. Wang, H. Yang, X. Huang, Prediction of the binding affinity of aptamers against the influenza virus, SAR and QSAR in Environmental Research, 30 (2019) 51–62.

[321] K. Khan, D. Baderna, C. Cappelli, C. Toma, A. Lombardo, K. Roy, E. Benfenati, Ecotoxicological QSAR modeling of organic compounds against fish: Application of fragment based descriptors in feature analysis, Aquatic Toxicology, 212 (2019) 162–174.

[322] K. Khan, K. Roy, Ecotoxicological risk assessment of organic compounds against various aquatic and terrestrial species: application of interspecies i-QSTTR and species sensitivity distribution techniques, Green Chemistry, 24 (2022) 2160–2178.

[323] G.J. Lavado, D. Baderna, D. Gadaleta, M. Ultre, K. Roy, E. Benfenati, Ecotoxicological QSAR modeling of the acute toxicity of organic compounds to the freshwater crustacean Thamnocephalus platyurus, Chemosphere, 280 (2021) 130652.

[324] T. Wang, L. Tang, F. Luan, M.N.D. Cordeiro, Prediction of the toxicity of binary mixtures by QSAR approach using the hypothetical descriptors, International Journal of Molecular Sciences, 19 (2018) 3423.

[325] Q. Jia, Y. Zhao, F. Yan, Q. Wang, QSAR model for predicting the toxicity of organic compounds to fathead minnow, Environmental Science and Pollution Research, 25 (2018) 35420–35428.

[326] Y. Huang, T. Li, S. Zheng, L. Fan, L. Su, Y. Zhao, H.-B. Xie, C. Li, QSAR modeling for the ozonation of diverse organic compounds in water, Science of the Total Environment, 715 (2020) 136816.

[327] X. Yu, Support vector machine-based model for toxicity of organic compounds against fish, Regulatory Toxicology and Pharmacology, 123 (2021) 104942.

[328] X. Yu, Prediction of chemical toxicity to Tetrahymena pyriformis with four-descriptor models, Ecotoxicology and Environmental Safety, 190 (2020) 110146.

[329] U. Maran, S. Sild, P. Mazzatorta, M. Casalegno, E. Benfenati, M. Romberg, Grid computing for the estimation of toxicity: acute toxicity on fathead minnow (Pimephales promelas), in: Distributed, High-Performance and Grid Computing in Computational Biology: International Workshop, GCCB 2006, Eilat, Israel, January 21, 2007. Proceedings, Springer, (2007), pp. 60–74.

[330] L.M. Su, X. Liu, Y. Wang, J.J. Li, X.H. Wang, L.X. Sheng, Y.H. Zhao, The discrimination of excess toxicity from baseline effect: Effect of bioconcentration, Science of the Total Environment, 484 (2014) 137–145.

[331] D. Gadaleta, A. Lombardo, C. Toma, E. Benfenati, A new semi-automated workflow for chemical data retrieval and quality checking for modeling applications, Journal of Cheminformatics, 10 (2018) 1–13.

[332] A. Mauri, V. Consonni, M. Pavan, R. Todeschini, Dragon software: An easy approach to molecular descriptor calculations, Match, 56 (2006) 237–248.

[333] S. Dimitrov, Y. Koleva, T.W. Schultz, J.D. Walker, O. Mekenyan, Interspecies quantitative structure-activity relationship model for aldehydes: aquatic toxicity, Environmental Toxicology and Chemistry: An International Journal, 23 (2004) 463–470.

[334] K.A. Hossain, K. Roy, Chemometric modeling of aquatic toxicity of contaminants of emerging concern (CECs) in Dugesia japonica and its interspecies correlation with daphnia and fish: QSTR and QSTTR approaches, Ecotoxicology and Environmental Safety, 166 (2018) 92–101.

[335] A. Toropov, E. Benfenati, QSAR modelling of aldehyde toxicity by means of optimisation of correlation weights of nearest neighbouring codes, Journal of Molecular Structure: THEOCHEM, 676 (2004) 165–169.

[336] W.H. Vaes, E.U. Ramos, H.J. Verhaar, J.L. Hermens, Acute toxicity of nonpolar versus polar narcosis: Is there a difference? Environmental Toxicology and Chemistry: An International Journal, 17 (1998) 1380–1384.

[337] K. Khan, K. Roy, Ecotoxicological modelling of cosmetics for aquatic organisms: a QSTR approach, SAR and QSAR in Environmental Research, 28 (2017) 567–594.

[338] N. Saha, F. Bhunia, A. Kaviraj, Toxicity of phenol to fish and aquatic ecosystems, Bulletin of Environmental Contamination and Toxicology, 63 (1999) 195–202.

[339] A. Levet, C. Bordes, Y. Clément, P. Mignon, H. Chermette, P. Marote, C. Cren-Olivé, P. Lantéri, Quantitative structure–activity relationship to predict acute fish toxicity of organic solvents, Chemosphere, 93 (2013) 1094–1103.

[340] T.M. Klapötke, Energetic Materials Encyclopedia, Walter de Gruyter GmbH & Co KG, (2018).

[341] M.H. keshavarz, Liquid Fuels as Jet Fuels and Propellants, Nova Science Publishers, New York, (2018).

[342] V. Singh, S. Panda, H. Kaur, P.K. Banipal, R.L. Gardas, T.S. Banipal, Solvation behavior of monosaccharides in aqueous protic ionic liquid solutions: Volumetric, calorimetric and NMR spectroscopic studies, Fluid Phase Equilibria, 421 (2016) 24–32.

[343] N. Abramenko, L. Kustov, L. Metelytsia, V. Kovalishyn, I. Tetko, W. Peijnenburg, A review of recent advances towards the development of QSAR models for toxicity assessment of ionic liquids, Journal of Hazardous Materials, 384 (2020) 121429.

[344] F. Yan, Q. Shang, S. Xia, Q. Wang, P. Ma, Topological study on the toxicity of ionic liquids on Vibrio fischeri by the quantitative structure–activity relationship method, Journal of Hazardous Materials, 286 (2015) 410–415.

[345] P. Luis, I. Ortiz, R. Aldaco, A. Irabien, A novel group contribution method in the development of a QSAR for predicting the toxicity (Vibrio fischeri EC50) of ionic liquids, Ecotoxicology and Environmental Safety, 67 (2007) 423–429.

[346] D.J. Couling, R.J. Bernot, K.M. Docherty, J.K. Dixon, E.J. Maginn, Assessing the factors responsible for ionic liquid toxicity to aquatic organisms via quantitative structure–property relationship modeling, Green Chemistry, 8 (2006) 82–90.

[347] R.N. Das, K. Roy, Development of classification and regression models for Vibrio fischeri toxicity of ionic liquids: green solvents for the future, Toxicology Research, 1 (2012) 186–195.

[348] M. Alvarez-Guerra, A. Irabien, Design of ionic liquids: an ecotoxicity (Vibrio fischeri) discrimination approach, Green Chemistry, 13 (2011) 1507–1516.

[349] S.P. Ventura, C.S. Marques, A.A. Rosatella, C.A. Afonso, F. Goncalves, J.A. Coutinho, Toxicity assessment of various ionic liquid families towards Vibrio fischeri marine bacteria, Ecotoxicology and Environmental Safety, 76 (2012) 162–168.

[350] S. Parvez, C. Venkataraman, S. Mukherji, A review on advantages of implementing luminescence inhibition test (Vibrio fischeri) for acute toxicity prediction of chemicals, Environment International, 32 (2006) 265–268.

[351] A. Fuentes, M. Lloréns, J. Saez, M.I. Aguilar, A.B. Pérez-Marín, J.F. Ortuño, V.F. Meseguer, Ecotoxicity, phytotoxicity and extractability of heavy metals from different stabilised sewage sludges, Environmental Pollution, 143 (2006) 355–360.

[352] M. Jafari, M.H. Keshavarz, H. Salek, A simple method for assessing chemical toxicity of ionic liquids on Vibrio fischeri through the structure of cations with specific anions, Ecotoxicology and Environmental Safety, 182 (2019) 109429.

[353] M.G. Montalbán, J.M. Hidalgo, M. Collado-González, F.G.D. Baños, G. Víllora, Assessing chemical toxicity of ionic liquids on Vibrio fischeri: correlation with structure and composition, Chemosphere, 155 (2016) 405–414.

[354] S.P. Costa, P.C. Pinto, R.A. Lapa, M.L.M. Saraiva, Toxicity assessment of ionic liquids with Vibrio fischeri: An alternative fully automated methodology, Journal of Hazardous Materials, 284 (2015) 136–142.

[355] H. Wang, S.V. Malhotra, A.J. Francis, Toxicity of various anions associated with methoxyethyl methyl imidazolium-based ionic liquids on Clostridium sp., Chemosphere, 82 (2011) 1597–1603.

[356] M. Montalbán, M. Collado-González, R. Trigo, F. Díaz Baños, G. Víllora, Experimental measurements of octanol-water partition coefficients of ionic liquids, Journal of Advanced Chemical Engineering, 5 (2015) 2.

[357] M.T. Garcia, N. Gathergood, P.J. Scammells, Biodegradable ionic liquids Part II. Effect of the anion and toxicology, Green Chemistry, 7 (2005) 9–14.

[358] X. Kang, Y. Zhao, H. Zhang, Z. Chen, Application of atomic electrostatic potential descriptors for predicting the eco-toxicity of ionic liquids towards leukemia rat cell line, Chemical Engineering Science, 260 (2022) 117941.

[359] N. Papaiconomou, J. Estager, Y. Traore, P. Bauduin, C. Bas, S. Legeai, S. Viboud, M. Draye, Synthesis, physicochemical properties, and toxicity data of new hydrophobic ionic liquids containing dimethylpyridinium and trimethylpyridinium cations, Journal of Chemical & Engineering Data, 55 (2010) 1971–1979.

About the Author

Mohammad Hossein Keshavarz (b. 1965), received a BSc in chemistry in 1988 from Shiraz University, Iran. He also received an MSc and a PhD at Shiraz University in 1991 and 1995, respectively. From 1997 until 2008, he was Assistant Professor, Associate Professor, and Professor of Physical Chemistry at the University of Malek Ashtar in Shahin Shahr, Iran. Since 1997, he has been Lecturer and researcher at the Malek Ashtar University of Technology, Iran. Keshavarz has published over 400 scientific papers in international peer-reviewed journals, five book chapters, and eight books in the field of assessment of energetic materials (four books in Persian and four books in English where three of them have second edition).

https://doi.org/10.1515/9783111189673-009

Index

https://doi.org/10.1515/9783111189673-010